An introduction to thermal physics

C. J. ADKINS

Lecturer in Physics in the University of Cambridge and Fellow of Jesus College, Cambridge

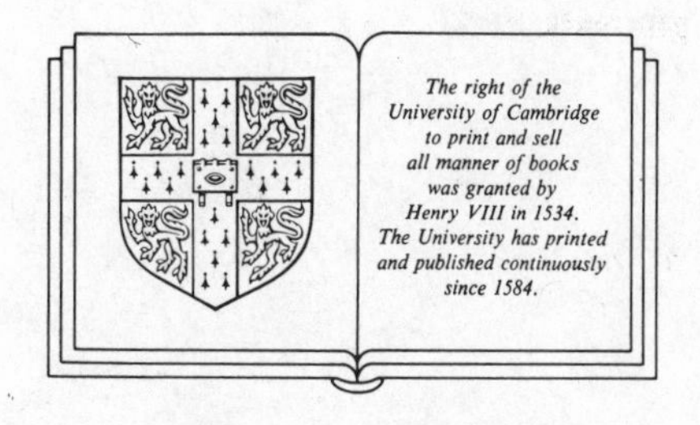

CAMBRIDGE UNIVERSITY PRESS

Cambridge

London New York New Rochelle

Melbourne Sydney

Published by the Press Syndicate of the University of Cambridge
The Pitt Building, Trumpington Street, Cambridge CB2 1RP
32 East 57th Street, New York, NY 10022, USA
10 Stamford Road, Oakleigh, Melbourne 3166, Australia

First published by Hodder and Stoughton as *Thermal physics* 1976
Reissued with amendments as *An introduction to thermal physics* by Cambridge University Press 1987

Printed in Great Britain by J. W. Arrowsmith Ltd, Bristol

British Library cataloguing in publication data

Adkins, C.J.
Introduction to thermal physics.

1. Thermodynamics
I. Title II. Adkins, C.J. Thermal physics
536'.7 QC311.15

Library of Congress cataloguing in publication data

Adkins, C. J. (Clement John)
An introduction to thermal physics.

Rev. ed. of: Thermal physics. 1976.
Includes index.
1. Heat. 2. Thermodynamics. 3. Gases, Kinetic theory of. I. Adkins, C. J. (Clement John) Thermal physics. II. Title.
QC254.2.A35 1987 536 86-17140

ISBN 0 521 33067 X hard covers
ISBN 0 521 33715 1 paperback

Contents

For Frances, Martin and Penelope

Preface

In recent years there have been marked changes in both the style and the content of physics courses at all levels. The general trend has been towards an increased emphasis on fundamental principles and microscopic explanations. As a consequence, the relative importance attached to various topics has changed, some new ones have been introduced and others, such as geometrical optics, virtually eliminated. Another consequence is that less detailed knowledge of numerous experimental techniques is now expected; in general, only a familiarity with the principles of various methods is required. These changes are reflected in revised syllabuses and call for a new generation of textbooks. It is in the spirit of these changes that this book is written.

The area covered corresponds very roughly to the traditional topics of *heat* and *kinetic theory* together with those parts of *properties of matter* for which there are simple explanations in terms of interatomic forces.

In level, the book is intended for use at universities and technical colleges in physics, engineering, chemistry or other science courses that require an elementary knowledge of thermal physics. It can be used in two ways: either as an introductory text, setting a firm foundation for further work in more specialized courses, or as an account, sufficient in itself for those requiring only a basic knowledge of the subject. It should also be useful in the science libraries of school sixth forms as a reference book that can help to point the way from school physics to the more mature approach of tertiary study. A familiarity with elementary calculus is assumed as its use is essential in the derivation of some of the fundamental results. However, when a result may also be derived without the use of calculus, an alternative derivation is generally given.

SI units are now essentially universal throughout science teaching and the text uses the system exclusively. It also generally follows the current recommendations of the Symbols Committee of the Royal Society as regards conventions for showing physical quantities and their units, and it incorporates the modern definitions of temperature scales.

The development of a sensibility towards the nature and magnitude of physical processes is an essential part of an education in physics and I consider the use of quantitative problems and exercises an essential means to this end. At the end of each chapter I have assembled a collection of problems. Many are original and some are inspired by ideas which have appeared in examination questions and in other texts. The problems are grouped according to the chapter subsections and, in each case, follow a brief summary of the key ideas of the relevant subsection. This makes it possible to work through the problems in parallel with the study of the text, so using them as an aid to learning. The arrangement is also useful for revision purposes. Generally I have avoided the inclusion of discursive questions, except where they are essential to cover certain key topics which cannot be tested quantitatively at that stage of the book. I have not included explicit worked examples separated out within the text. There are several reasons for this, among which is the fact that numerical examples are worked out as part of the narrative, and also, in many cases, the quantitative development of a key result serves to illustrate how that key result may be used.

Despite its title, this is not an easy book, for although it only deals with the *foundations* of thermal physics, the text challenges the reader to think deeply and carefully about the concepts and methods of the subject. It also seeks to show the relevance of the subject by relating these concepts and methods to the everyday physical world. Thus, although a simplistic reading will yield some understanding of thermal physics, more thoughtful study will bring extra rewards.

I should like to thank my wife, Tessara, for her suport and encouragement while I was writing this book, and to thank her and Jean Millar for their help with some of the illustrations. I am also most grateful to C. B. Spurgin whose advice and criticisms as the book took shape were much appreciated.

It is also a pleasure to thank: Dr J. Ashmead and the Institute of Physics for permission to use the photograph reproduced in figure 4.15; Dr A. M. Glazer for supplying me with the X-ray scattering picture from which figure 3.5 was prepared; AGA Infrared Systems AB for supplying me with the thermogram reproduced in figure 6.7; Foster Cambridge Ltd. for the disappearing filament illustration of figure 1.22; RS Components Ltd. for permission to reproduce the thermistor data shown in figure 1.19; and the Escher Foundation at Haags Gemeente-museum, The Hague, for permission to reproduce Escher's *Waterfall* in figure 2.4 and on the cover.

My objective throughout the detailed writing of this book has been to achieve a clear and stimulating exposition: to write a book which is easy to learn from. Those who use it must judge whether I have been successful.

C. J. Adkins
Cambridge, 1986

Units, Symbols and Conventions

This book uses SI (Système International) units and generally follows the current recommendations of the Symbols Committee of the Royal Society as regards symbols and conventions of notation.*

The names and symbols for the SI base units are:

Physical quantity	*Name of SI unit*	*Symbol for SI unit*
length	metre	m
mass	kilogram	kg
time	second	s
electric current	ampere	A
thermodynamic temperature	kelvin	K
luminous intensity	candela	cd
amount of substance	mole	mol

The SI units of certain common physical quantities have special names. Some of those used in this book are listed in the table below.

Physical quantity	*Name of SI unit*	*Symbol for SI unit*	*Definition of SI unit*	*Equivalent forms*
force	newton	N	$\mathrm{m\,kg\,s^{-2}}$	$\mathrm{J\,m^{-1}}$
energy	joule	J	$\mathrm{m^2\,kg\,s^{-2}}$	$\mathrm{N\,m}$
pressure	pascal	Pa	$\mathrm{m^{-1}\,kg\,s^{-2}}$	$\mathrm{N\,m^{-2}}$, $\mathrm{J\,m^{-3}}$
power	watt	W	$\mathrm{m^2\,kg\,s^{-3}}$	$\mathrm{J\,s^{-1}}$
electric charge	coulomb	C	$\mathrm{s\,A}$	$\mathrm{A\,s}$
electric potential difference	volt	V	$\mathrm{m^2\,kg\,s^{-3}\,A^{-1}}$	$\mathrm{J\,A^{-1}\,s^{-1}}$, $\mathrm{J\,C^{-1}}$
electric resistance	ohm	Ω	$\mathrm{m^2\,kg\,s^{-3}\,A^{-2}}$	$\mathrm{V\,A^{-1}}$
electric capacitance	farad	F	$\mathrm{m^{-2}\,kg^{-1}\,s^4\,A^2}$	$\mathrm{A\,s\,V^{-1}}$, $\mathrm{C\,V^{-1}}$
inductance	henry	H	$\mathrm{m^2\,kg\,s^{-2}\,A^{-2}}$	$\mathrm{V\,A^{-1}\,s}$
frequency	hertz	Hz	$\mathrm{s^{-1}}$	

Angles, though formally defined so as to be dimensionless (see Appendix, section A.1), are sometimes considered as supplementary units:

Physical quantity	*Name of SI unit*	*Symbol for SI unit*
plane angle	radian	rad
solid angle	steradian	sr

The International System has a set of prefixes which may be used to construct decimal multiples of units.†

Multiple	*Prefix*	*Symbol*	*Multiple*	*Prefix*	*Symbol*
10^{-1}	deci	d	10	deca	da
10^{-2}	centi	c	10^2	hecto	h
10^{-3}	milli	m	10^3	kilo	k
10^{-6}	micro	μ	10^6	mega	M
10^{-9}	nano	n	10^9	giga	G
10^{-12}	pico	p	10^{12}	tera	T
10^{-15}	femto	f			
10^{-18}	atto	a			

* *Quantities, Units and Symbols*, The Royal Society (London, 1975).

† μ, the prefix meaning one millionth, is the Greek letter *mu*.

Symbols for units are always printed in roman (upright) type while symbols for physical quantities such as p for pressure are printed in italic (sloping) type.

The value of a physical quantity is always equal to the product of a numerical value and a unit. Thus, the physical quantity, the mass m_e of the electron is given by

$$m_e = 9.11 \times 10^{-31}\ \mathrm{kg}$$

In this equation, the magnitude and unit of the physical quantity are equated across the equals sign. An equivalent dimensionless equation is

$$m_e/\mathrm{kg} = 9.11 \times 10^{-31}$$

Here, the solidus (/) on the left hand side represents division in the usual way: the physical quantity (mass of an electron) is divided by a unit of mass (the kilogram) and the result is a pure number, 9.11×10^{-31}, the number of kilograms in the mass of one electron.

The use of the solidus to represent division of a physical quantity by a unit may be applied in several other ways.

a) *More complex equations* relating physical quantities may often be written concisely and unambiguously using the solidus notation. For example, the molar heat capacity at constant pressure of copper C_{pm} depends on thermodynamic temperature T at low temperatures according to the equation

$$C_{pm}/\mathrm{kJ\ K^{-1}\ mol^{-1}} = 1.94\,(T/\Theta)^3$$

where $\Theta = 348$ K is the 'Debye Temperature' of copper. Here, $\mathrm{kJ\ K^{-1}\ mol^{-1}}$ is the unit in which the heat capacity is measured so that the left hand side is a dimensionless number. Similarly, the term in brackets on the right is dimensionless because both T and Θ are temperatures measured in kelvins. Clearly, the number 1.94 is also dimensionless to that the equation is dimensionally homogeneous throughout. An alternative way of giving the same information would be to write

$$C_{pm} = a(T/\Theta)^3$$

where

$$a = 1.94\ \mathrm{J\ K^{-1}\ mol^{-1}}$$

In this form, dimensional quantities are equated across the equals sign. But it is *wrong* to write

$$C_{pm} = 1.94\,(T/\Theta)^3$$

for this equation is dimensionally inconsistent: the left side has dimensions of heat capacity while the right side is dimensionless. The equation only 'works' if C_{pm} is measured in the right units. An equation should express a physical fact; and since a true physical fact is true regardless of how (in what units) it is measured, this last form is unacceptable.

b) *In tables and graphs* the numbers entered or plotted are dimensionless so that the labelling of the table headings or graph axes should also be dimensionless. Thus, a graph of pressure against temperature might have its axes labelled 'pressure/mmHg' and 'temperature/K' respectively. The result of dividing the physical quantity pressure by the unit of pressure mmHg is a pure number, and so is the result of dividing temperature by kelvins. The axes are therefore calibrated in pure numbers and it is the relationship between two pure numbers which the graph displays. It is *not* correct to label the axes 'pressure (mmHg)' and 'temperature (K)' because, following normal notation, these would mean either pressure multiplied by mmHg and temperature multiplied by kelvins, or pressure, a function of the unit mmHg, and temperature, a function of the kelvin. Either alternative is nonsense. Nevertheless, such forms have been used in the past with the meaning 'the numbers on this axis give the magnitude of the pressure when it is measured in mmHg', etc. Clearly, there is no need to have to adopt a special meaning to the use of brackets when the solidus notation is physically and mathematically correct and totally unambiguous.

c) The solidus notation is also useful for *changing the units in which a physical quantity is measured.* It reduces conversion of units to routine algebra. Suppose the speed u of a car is $90\ \mathrm{km\ h^{-1}}$ and we wish to find its speed in $\mathrm{m\ s^{-1}}$. We are given $u/\mathrm{km\ h^{-1}} = 90$ and we want $u/\mathrm{m\ s^{-1}}$. Following the normal rules of algebra we may write

$$u/\mathrm{m\ s^{-1}} = (u/\mathrm{km\ h^{-1}}) \times (\mathrm{km/m}) \times (\mathrm{s/h})$$

The first bracket on the right is the number given, the second is the pure number which results from dividing 1 km by 1 m, namely 1000; and the third term on the right is the pure number which results from dividing the unit of time, 1 s by the unit of time 1 h, namely 1/3600. Thus,

$$u/\mathrm{m\ s^{-1}} = 90 \times 1000 \times 1/3600 = 25$$

or

$$u = 25\ \mathrm{m\ s^{-1}}$$

Finally, we list the symbols used in this book.

ROMAN LETTERS

a	constants
b	constants
c	molecular speed
d	distance
	infinitesimally small change in
e	spectral emissive power
	electronic charge
	base of natural logarithms
f	number of degrees of freedom
	a function
g	acceleration of free fall
h	Planck constant
i	an integer
k	Boltzmann constant
l	mean free path
m	mass
n	number of moles
	number density
p	pressure
r	radius, distance
t	Celsius temperature
	time
x	a variable
y	a variable
z	coordination number
	a variable
A	area
C	heat capacity
	capacitance
E	Young modulus
	electromotive force
	electric field strength
F	force
H	scale height
I	current
J	joule
	current density
K	kelvin
	bulk modulus
L	length
	latent heat
	inductance
	Lorenz number
M	molar mass
P	power
	probability
Q	heat
	charge
R	resistance
	molar gas constant
S	entropy
	molecular diameter
T	thermodynamic temperature
U	internal energy
V	volume
	potential difference
$\mathscr{V}$	speed
W	work

GREEK LETTERS

Letter	*Name*	*Meanings*
α	alpha	spring constant, linear expansivity, absorptivity
β	beta	cubic expansivity
γ	gamma	ratio of principal heat capacities, surface tension
δ	delta	small change in
Δ	delta (cap.)	finite increment of
ε	epsilon	energy, emissivity
η	eta	viscosity, efficiency
θ	theta	angle
Θ	theta (cap.)	empirical temperature
κ	kappa	compressibility
λ	lambda	thermal conductivity, wavelength
μ	mu	one millionth
ν	nu	frequency
π	pi	ratio of circumference to diameter of circle
ρ	rho	density
σ	sigma	electrical conductivity, Stefan–Boltzmann constant
ϕ	phi	potential
Φ	phi (cap.)	flux density
ω	omega	angular frequency
Ω	omega (cap.)	solid angle, ohm

MATHEMATICAL NOTATION

$+$	plus
$-$	minus
$=$	equal to
$\neq$	not equal to
$\approx$	approximately equal to
$\propto$	proportional to
$<$	smaller than

$>$	larger than
$\leqslant$	smaller than or equal to
$\geqslant$	larger than or equal to
$\ll$	much smaller than
$\gg$	much larger than
$\langle a \rangle, \bar{a}$	mean value of a
$f(x)$	function of x
$\lim_{x \to a} f(x)$	the limit to which $f(x)$ tends as x approaches a
Δ	finite increment of
δ	small change of
$\dfrac{\mathrm{d}f}{\mathrm{d}x}$	differential coefficient of $f(x)$ with respect to x
$\left(\dfrac{\partial f}{\partial x}\right)_y$	partial differential coefficient of $f(x, y)$ with respect to x when y is held constant
$\mathrm{d}f$	total differential of f (infinitesimal change in f)
$\int f(x)\,\mathrm{d}x$	the integral of $f(x)$ with respect to x
$\oint f(x)\,\mathrm{d}x$	the integral of $f(x)$ with respect to x around a closed path
e^x, $\exp x$	exponential of x
e	base of natural logarithms
$\ln x$	natural logarithm (logarithm to the base e) of x
$\lg x$	common logarithm (logarithm to the base 10) of x

1 Temperature

1.1 INTRODUCTION

Thermal physics is the study of those properties of materials which are affected by temperature. It is an enormous field, having something to say about subjects as diverse as the expansion of a solid, the internal constitution of stars and why the electrical resistance of some metals vanishes at low temperatures.

The job of the scientist is to observe nature, to try to recognize the regularities in its behaviour and to seek to link those regularities together by laws and principles. For example, it is often found that the current flowing in an electrical conductor is proportional to the applied potential difference; this regularity is called Ohm's law.

When the scientist tries to 'explain' his observations, there are two levels of explanation he can attempt. He may content himself with ideas relating to the behaviour of matter in bulk (e.g. Ohm's law). This is the *macroscopic* approach. In the area of thermal physics it leads to the subject called *thermodynamics.* This branch of physics developed most rapidly during the last century in connection with the study of machines, such as the steam engine, which supplied power for the new industries; but, it was soon realized that the laws of thermodynamics were very fundamental and of importance in areas quite different from power engineering where they were developed. The laws of thermodynamics provide a theoretical framework which is used in many branches of modern science. A typical result which may be derived by thermodynamics is the Clausius–Clapeyron equation (page 111) which connects the variation with temperature of vapour pressure with latent heat.

At the other level of explanation, the scientist tries to base his understanding on ideas of the nature and properties of matter at the atomic level. This is *microscopic* physics. An example is the kinetic theory of gases (chapter 3) which 'explains' the bulk behaviour of gases in terms of the properties of the molecules of which the gas is composed. While models and laws relating to matter at the atomic level play no direct part in macroscopic physics, they are the essence of microscopic physics. In thermal physics, the microscopic approach leads to the subject called *statistical mechanics.* A typical result of the arguments of statistical mechanics is the Maxwell distribution (page 54), which gives the probabilities of different molecular speeds in a gas.

Both macroscopic and microscopic approaches are of value, as we shall see later in this book, for there are times when we need to explain in terms of fundamental laws operating at the atomic level, while there are other times when to involve ourselves with detailed microscopic models would be a positive encumbrance.

1.2 SOME BASIC IDEAS

We call the object we are investigating the *system.* It could be a volume of gas, or a spring, or a refrigerator, or a solid in the process of melting. We shall often illustrate our arguments by taking as a model system a given mass of gas in a cylinder with a frictionless piston. This is a good model to take because it is easy to visualize what happens when we heat it or do work on it by compressing it.

We describe the *state* of the system in terms of appropriate *parameters* or *variables,* such as mass, pressure, volume, density, temperature. When a system is in a given state, we will always get the same results for any measurements we may make on it. The variables are not all independent; some are related to one another. For example, density is mass/volume. For simple systems of given mass we find that we need to fix the values of *two* variables in order to fix the state of the system. For our given mass of gas, for example, we find that, if we first set the volume, we may still adjust the pressure to any value we please (by varying the temperature), but once the pressure is set also, there is no other property which can be varied. We therefore say that the gas has two *degrees of freedom*: we are free to choose two of the variables as we like, but then all the other parameters will have taken up definite values which we cannot adjust. This means that any mathematical equation connecting system variables must have at least three variables in it: the values of two must be known in order to determine the state of the system and so fix the value of a third. Thus, the equation of state of an ideal gas (page 40) connects pressure, volume and temperature: $pV/T =$ constant. Again, this is why we have two common heat capacities. A heat capacity is the rate at which heat is absorbed as we change the temperature (section 2.3); but since simple systems have two degrees of freedom, we do not know *how* the system is going to change as the temperature is changed unless more information is given. Two simple cases are that the system should be kept at constant volume or at constant pressure. Such a condition is called a *constraint* because it constrains the system to change in a certain way. Any constraint removes the second degree of freedom so that now a definite amount of heat is absorbed as the temperature is changed.

In developing the ideas of temperature, heat, and so on, we shall be interested in the ways systems interact with one another. There are two kinds of interaction. In *work-like* interactions, one system does work on another (figure 1.1): a force moves through a given distance as in compressing a gas or stretching a spring, or a battery will do work in charging a capacitor because it forces charge to flow against an opposing potential difference. The other kind of interaction is *thermal* and is typified by flow of heat when we place a hot body in contact with a cold one (figure 1.2). In both kinds of interaction,

Figure 1.1 A work-like interaction

Figure 1.2 A thermal interaction

energy is transferred from one system to another; the difference is that in work-like interactions the process involves some sort of large scale motion (all the molecules of the piston move forward together as the gas is compressed), whereas in thermal interactions the energy is associated with disordered thermal motions of the atoms of which the system is composed: the random motions of the molecules of a gas or the vibrations of the atoms of a solid. When a thermal interaction takes place, the atoms of the cooler system are excited into more violent thermal motion by contact with the hotter (more energetic) atoms of the hotter system.

If two systems are placed in contact in such a way as to allow thermal interaction, they are said to be in *thermal contact* (figure 1.3). If a system is prevented from interacting thermally with its surroundings it is said to be *thermally isolated* (coffee in a vacuum flask), and any change it undergoes is said to be an *adiabatic* change. Put differently, an adiabatic change is one in which no heat enters or leaves the system. We shall discuss the ideas of heat and work more fully in chapter 2.

If you have found these introductory ideas confusing at this stage, it is probably because I have been giving a kind of preview of the things we shall be discussing carefully later in the book. Perhaps you should return to sections 1.1 and 1.2 when you have finished reading it!

Figure 1.3 Two systems making thermal contact

1.3 TEMPERATURE

The idea of temperature almost certainly originally arose from the physiological sensation of hotness, an unreliable measure of temperature (figure 1.4). We can, however, develop a more exact concept of what temperature means by discussing what happens when bodies are placed in thermal contact.

If we take two systems and place them in thermal contact we generally find that initially changes will occur in both (figure 1.5). Eventually the changes cease, and the systems are then said to be in *thermal equilibrium.* We introduce the idea of temperature by saying that *the condition for the systems to be in thermal equilibrium is that they should be at the same temperature.* Conversely, two systems which are already at the same temperature will not undergo change when placed in thermal contact. Equality of temperature is therefore the condition for thermal equilibrium.

However, defined like this, it is not clear that the condition for thermal equilibrium might not depend on the nature of the systems concerned. Can the same condition apply when we put a thermometer in our mouth to take our temperature as applies when we put a thermometer in a beaker of concentrated sulphuric acid? Experience shows that the condition for thermal equilibrium does not depend on the nature of the systems concerned. This experimental fact is embodied in the *zeroth law of thermodynamics* (so called because the need for it was not recognized until after the first law had been established).

If two systems are separately in thermal equilibrium with a third, then they must also be in thermal equilibrium with each other.

Since we are free to choose anything we like for the third system, it follows that the condition for thermal equilibrium cannot depend on the nature of the systems concerned: all systems in thermal equilibrium have the same temperature irrespective of their nature. Another way of expressing this is to say that temperature is a universal property. The kind of experiment with which we might illustrate the zeroth law is shown in figure 1.6.

We should note that if we require one system to be in thermal equilibrium with another (i.e. at the same temperature) this represents a constraint on the system and removes one of its degrees of freedom. With our fixed mass of gas, for example, in the absence of any constraint, we are free to choose pressure and volume as we please; however, as soon as we require thermal equilibrium with some other system, the temperature is fixed, and for every value of volume there is only one possible value of pressure. We can put this another way by saying that for each temperature there is a

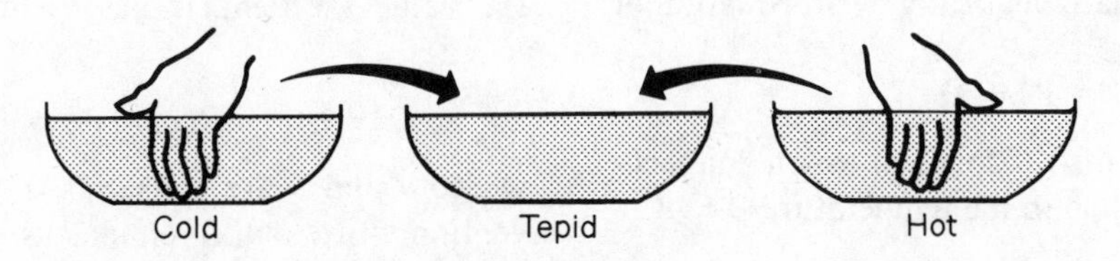

Figure 1.4 The physiological sensation of hotness is an unreliable measure of temperature. After one's hand has been in cold water for a time, tepid water feels hot. After it has been in hot water, the tepid water feels cold.

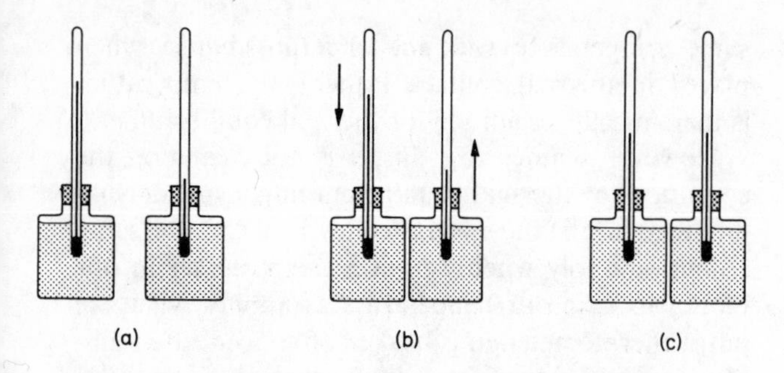

Figure 1.5 When two systems (a), are placed in thermal contact (b), changes generally take place in both systems until they reach thermal equilibrium (c). Here, thermometers indicate the changes (of temperature).

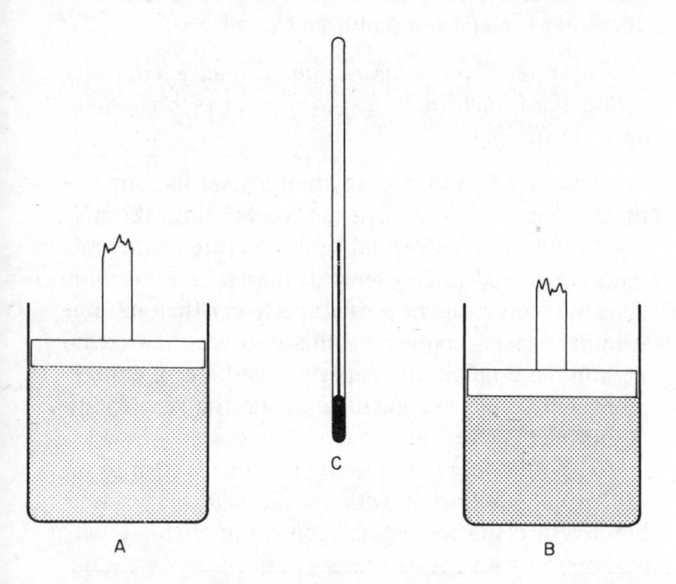

Figure 1.6 An illustration of the zeroth law. Systems A and B are fixed masses of gases at particular values of pressure and volume. The third system, C, is a thermometer. If C is in thermal equilibrium separately with A and B, so that its reading does not change when it is placed in thermal contact with either, then no changes will take place when A and B are placed in thermal contact with each other (because they are already at the same temperature).

unique relationship between pressure and volume. This is represented mathematically by an equation of the form

$$f(p, V) = \Theta$$

where $f(p, V)$ stands for the formula which connects the pressure and volume to the temperature Θ.*

* The Greek letter *theta*. For temperature we use the capital letter. The lower case letter, θ, is generally used for angles.

For an ideal gas, f is a simple product, pV, and for constant temperature we have

$$f(p, V) = pV = \Theta = \text{constant}$$

which is Boyle's law (page 40). The curve relating p to V for a given value of Θ is called an *isotherm* (figure 1.7).

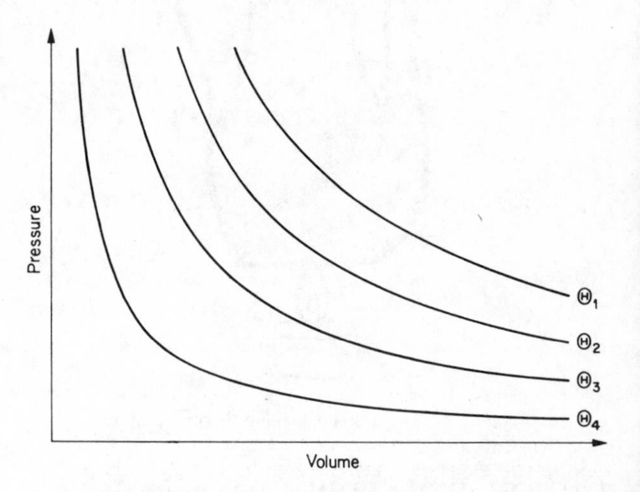

Figure 1.7 Isotherms for an ideal gas. For each temperature, there is only one possible value of pressure for each value of volume.

1.4 SCALES OF TEMPERATURE

We have introduced temperature as a rather abstract concept connected with thermal equilibrium. In practice, we would like to be able to represent temperature by a number whose magnitude changes in some regular way in relation to our ideas of hotness: the hotter the body, the larger the number representing temperature. This is what we do when we set up a *scale of temperature*. The easiest way to do this is to choose a convenient system with a property x which changes with temperature and take the value for temperature as linearly proportional to x:

$$\Theta(x) = ax + b \qquad (1.1)$$

where a and b are constants. Here, again, the convention written $\Theta(x)$ reminds us that Θ is a quantity whose value depends on the value of x: Θ is said to be a *function* of x. The relationship of equation *1.1* is illustrated in figure 1.8.

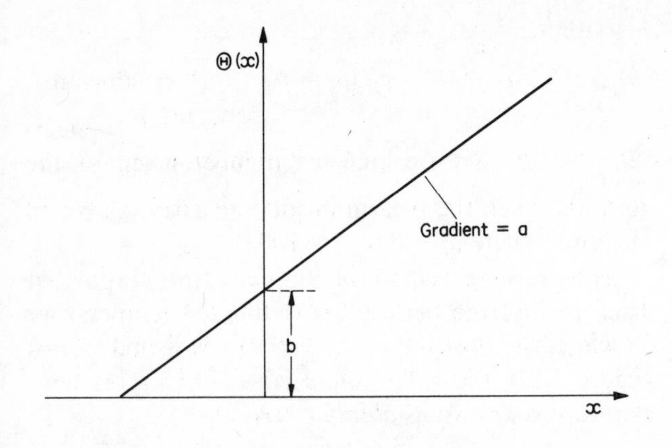

Figure 1.8 A plot of the equation $\Theta(x) = ax + b$

When we set up a temperature scale for everyday use, we want a conveniently placed zero, say low in the range of commonly encountered temperatures, and a sensibly sized unit. Conventionally, the *ice point*, the temperature at which water freezes at one atmosphere pressure, is taken as 0°, and the *steam point*, the temperature at which water boils at one atmosphere pressure, is taken as 100°. A scale so constructed is known as a *centigrade* scale. (Centigrade means one hundred steps.) We may take some examples.

A temperature on a centigrade scale based on the expansion of mercury in a mercury in glass thermometer uses the length of mercury in the capillary as the *thermometric* (temperature measuring) quantity x. The centigrade temperature is given by

$$\Theta(L) = (L - L_i) \times \frac{100}{(L_s - L_i)} \qquad (1.2)$$

where L, L_s and L_i are the lengths of the mercury at the temperature to be measured and at the ice and steam points respectively (figure 1.9). Comparing

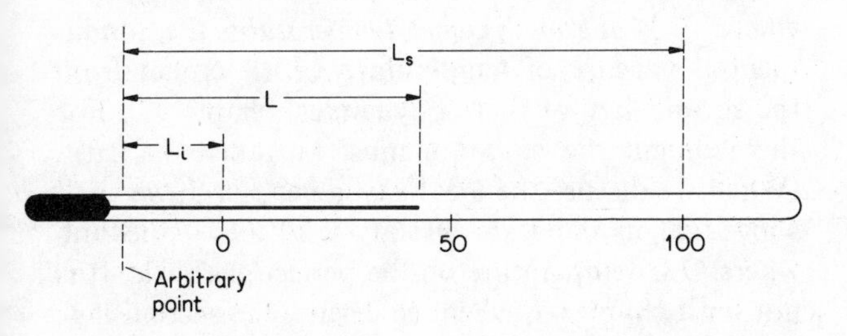

Figure 1.9 Constructing a centigrade scale for a mercury-in-glass thermometer

equation *1.2* with the general linear form, equation *1.1*, we find

$$a = 100/(L_s - L_i)$$
$$b = -100L_i/(L_s - L_i)$$

Again, a temperature on a centigrade scale based on the variation of the resistance R of a coil of copper wire is given by

$$\Theta(R) = 100(R - R_i)/(R_s - R_i) \qquad (1.3)$$

where R, R_i and R_s are the values of the resistance at the temperature to be measured, and at the ice and steam points respectively.

Scales constructed in this way are known as *empirical* scales: empirical means based on experiment. The symbol Θ is always used for empirical temperature.

When the scales are set up like this, we find that the value obtained for a temperature depends on what thermometer we use. This is because different properties do not respond in the same way to change of temperature. As a result, centigrade scales based on different systems will not generally agree with one another except, of course, at 0° and 100°, the calibration points, where they must agree by definition. Figure 1.10 shows the differences between centigrade temperatures determined with different thermometers.

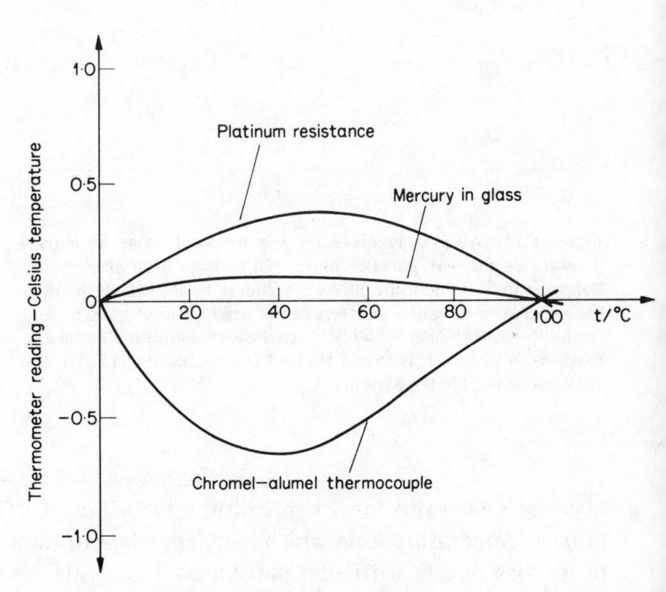

Figure 1.10 Differences between centigrade scales of common thermometers over the temperature range 0–100 °C

In the search for a scale of temperature which did not depend on the properties of particular substances it was found that disagreement was small among thermometers based on the behaviour of gases. Gases have two degrees of freedom, so that a constraint has to be applied if the pressure or volume is to be uniquely related to temperature. Commonly, volume is kept constant and pressure used as the thermometric property. This is a *constant volume gas thermometer* (page 8) and will give a centigrade temperature

$$\Theta(p) = 100(p - p_i)/(p_s - p_i) \qquad (V = \text{constant})$$

Now while disagreements are generally small when temperatures are determined with different gases at normal pressures in this way, it is found that the differences which are present become smaller as the pressures used are reduced. If the measurements are *extrapolated* to find the temperature which would be given if the pressure could be reduced to zero (figure 1.11), it is found that, in this limit, *all* gas thermometers

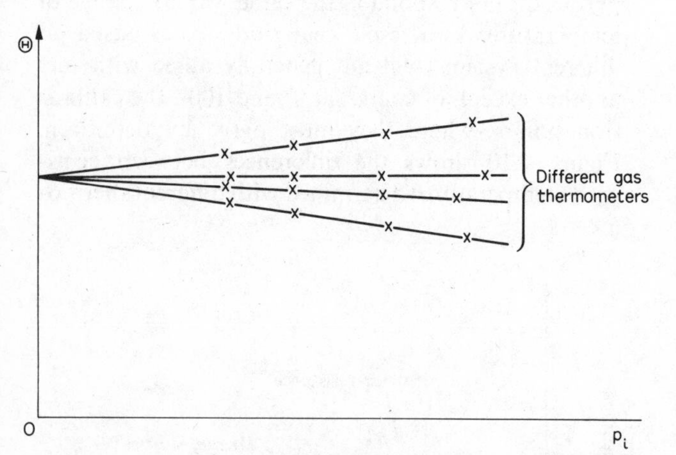

Figure 1.11 All gas thermometers give the same value for a temperature in the low pressure limit. With each thermometer, the temperature is found using different values of p_i. If the lines through the experimental values of temperature are extended back to find the temperature which would be given if the measurements could be made with $p_i = 0$, it is found that all the thermometers give the same value for the temperature.

give the *same* value for a temperature. One thus obtains a temperature scale which is independent of the properties of any particular substance. This scale is called the *perfect gas centigrade scale*. On the perfect gas centigrade scale, the expression for the temperature is written

$$\Theta_{pgc} = \lim_{p_i \to 0} 100(p - p_i)/(p_s - p_i) \qquad (V = \text{constant}) \qquad (1.3)$$

$\lim_{p_i \to 0}$ means that we take the number given by the formula when the measurements are extrapolated to the limit where $p_i = 0$ (figure 1.11).

If measurements made on this scale are extrapolated back to find the perfect gas centigrade temperature at which the pressure of the perfect gas would vanish (figure 1.12), the value found is -273.15. This temperature is known as *absolute zero*.

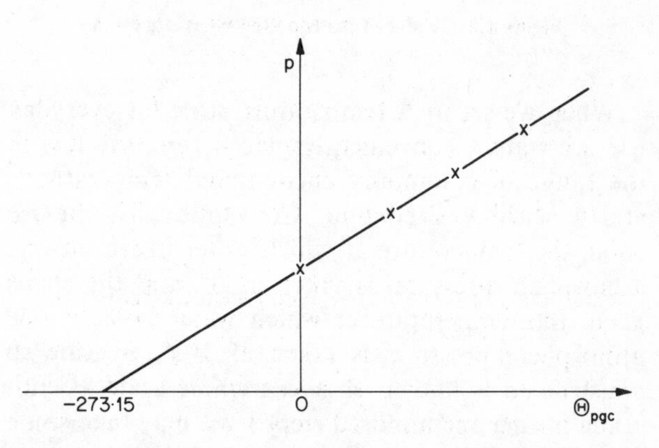

Figure 1.12 The pressure of an ideal gas would vanish at a perfect gas centigrade temperature of −273.15

1.5 THERMODYNAMIC TEMPERATURE

The reason why all gases give the same value for a temperature in the low pressure limit is that, in that limit, their behaviour tends to that of the perfect or ideal gas (chapter 3) whose equation of state is

$$pV/T = \text{constant} \qquad (1.4)$$

where T is *thermodynamic temperature*, the fundamental measure of temperature which comes from the second law of thermodynamics (chapter 7). For the moment the equation must be taken on trust. (When we discuss the ideal gas in chapter 3, we shall show that its equation of state is $pV/\Theta = $ constant, where Θ is temperature on the perfect gas scale. It is not until chapter 7, when we discuss the second law, that we are able to explain the idea of thermodynamic temperature and prove that perfect gas temperature is identical to thermodynamic temperature.) Substituting *1.4* in *1.3*, the volume and the constant cancel through numerator and denominator and we get

$$\Theta_{pgc} = (T - T_i) \times \frac{100}{(T_s - T_i)}. \qquad (1.5)$$

This equation shows that perfect gas centigrade temperatures, like thermodynamic temperatures, are independent of the properties of any particular substance.

Thermodynamic temperature is now accepted as the fundamental measure of temperature. Gases are unique in that thermodynamic temperature appears in such a simple way in the equation of state (which holds for real gases in the low pressure limit), and this is why determinations of thermodynamic temperatures are often ultimately based on gas thermometry.

Now we are free to choose the size of the unit of thermodynamic temperature as we please. If we choose to have 100 units between ice and steam points, equation *1.5* becomes

$$\Theta_{pgc} = (T - T_i)/\mathrm{K} \qquad (1.6)$$

However, there are disadvantages in fixing the size of the unit in this way. In the first place it is necessary to calibrate a gas thermometer at *two* fixed points: the ice and steam points. Secondly, when measurements made at the fixed points are extrapolated back to very low temperatures, that is to small values of T, the experimental uncertainties become *relatively* large, which could be serious in low temperature work. Now we notice that thermodynamic temperature has a natural zero, namely, the temperature at which the pressure of an ideal gas would vanish (see equation *1.4*). If we take this natural zero as a fixed point on the scale, we only need to calibrate the thermometer at *one* fixed point to fix the size of the unit. In effect, we choose the *value* of thermodynamic temperature for *one* fixed point and this sets the scale. This way of fixing the size of the unit of thermodynamic temperature is the one now adopted by the International Committee of Weights and Measures. The fixed point chosen is the *triple point of water* (the temperature at which water, ice and water vapour coexist in equilibrium, a temperature more reproducable than the ice or steam points) and the value of temperature allotted to it is 273.16. The unit so defined is called the kelvin* for which the symbol K is used. Thus

Thermodynamic temperature is the fundamental temperature; its unit is the kelvin which is defined as the fraction 1/273.16 of the thermodynamic temperature of the triple point of water.

Therefore, a thermodynamic temperature determined by gas thermometry would be given by

$$T/\mathrm{K} = \lim_{p \to 0} (pV) \times \frac{273.16}{\lim_{p \to 0} (pV)_{tr}} \qquad (1.7)$$

where pV is the value of the product of pressure and volume at the temperature to be determined and $(pV)_{tr}$ is the value of the product at the triple point. Absolute zero is 0 K by definition. When we discuss the ideal gas we shall see that absolute zero is the temperature at which all thermal motion would cease.

Figure 1.13 shows schematically how the triple point is achieved for thermometer calibration.

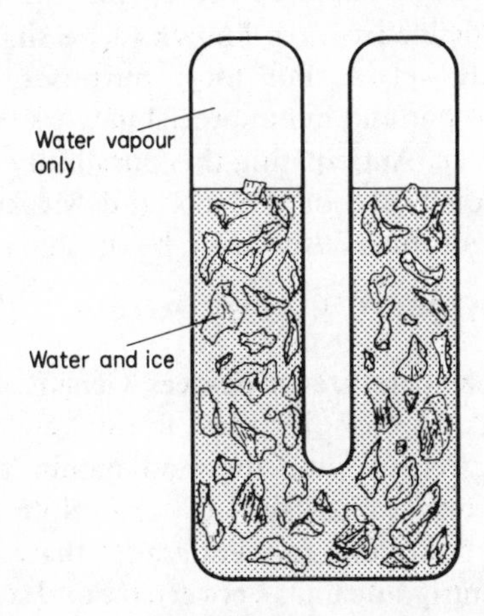

Figure 1.13 A triple point cell. The cell is cooled until some ice is present as well as water and water vapour. When all three are in equilibrium with one another, the temperature is 273.16 K by definition. The thermometer to be calibrated is inserted into the central tube.

1.6 THE CELSIUS TEMPERATURE SCALE

The reason for the choice of 273.16 K for the thermodynamic temperature of the triple point of water

* After Lord Kelvin (William Thompson), 1824–1907.

was that it gave *to the accuracy of the best measurements then available* exactly 100 K between ice and steam points. This meant that, within experimental error, the perfect gas centigrade scale and thermodynamic temperature coincided precisely apart from the shift of zero to the ice point:

$$\theta_{pgc} = T/\text{K} - 273.15 \qquad (1.8)$$

This is the same as *1.6* but with the value of T_i substituted. However, more recent measurements have revealed previously unsuspected errors in conventional gas thermometry. In particular, it has been shown that, to a precision better than 0.01 K, $T_s - T_i = 99.97$ K. We therefore now have:

$$\begin{aligned} \text{triple point} &= 273.16\ \text{K} \quad \text{by definition} \\ \left.\begin{aligned} \text{ice point} &= 273.15\ \text{K} \\ \text{steam point} &= 373.12\ \text{K} \end{aligned}\right\} &\ \text{by experiment} \end{aligned}$$

Equation *1.5* still defines a perfect gas *centigrade* scale, of course, because the term $100/(T_s - T_i)$ takes account of the improved value for T_i; but the centigrade unit so defined is now known to be slightly smaller than the kelvin. For most purposes, the difference is unimportant, but in precision work there could be confusion. Anticipating this possibility, the International Committee of Weights and Measures defined a *new* scale, the *Celsius scale*, by the equation

$$t/°\text{C} = T/\text{K} - 273.15 \qquad (1.9)$$

where t is the temperature in degrees Celsius. The Celsius degree is therefore identical to the kelvin *by definition*, and Celsius and thermodynamic temperatures differ only by the shift of zero. Note that Celsius is not a new name for centigrade: there will always be 100 centrigrade units between ice and steam points because that is how a centigrade scale is defined, but experiment shows the temperature differences to be significantly less than 100 kelvins.

The temperatures of the primary reference points given in table 1.1 are the values adopted to define the International Practical Temperature Scale of 1968 (see section 1.8). In the next few years, a new set of temperatures will be allotted to the reference points and a new practical scale will be defined. The maximum correction will be less than 0.5 K.

Note that the symbol t is always used for Celsius temperatures and, whereas no degree sign is used with K for the kelvin, the degree sign is included with the C for Celsius temperatures: °C.

1.7 SOME COMMON THERMOMETERS

In choosing a thermometer for a particular application the important criteria are

accuracy, the ability to give an accurate value for the thermodynamic temperature,
sensitivity, the ability to measure very small changes of temperature,
suitability, satisying special requirements like small size, rapid response, robustness, resistance to corrosion, ability to operate remote from the observer,
convenience, simplicity of construction and operation.

The best choice in any particular case depends on the task in hand.

Gas thermometers are important because they have been widely used for basic determinations of thermodynamic temperature; but they are inconvenient, and, in order to obtain a good degree of accuracy, elaborate precautions have to be taken and complicated corrections made. The essential elements of a simple constant volume gas thermometer are shown in figure 1.14. The bulb *B* is immersed in the region whose temperature is to be measured. The tube connecting the bulb to the manometer is usually capillary so that the amount of gas not at the temperature being measured is relatively small. For the same reason, the bulb is usually relatively large (in

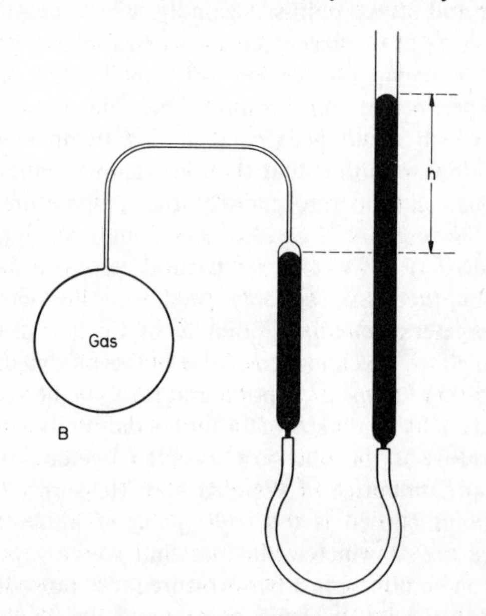

Figure 1.14 The elements of a simple constant volume gas thermometer

Table 1.1 Temperatures of the standard and primary reference points as adopted for the International Practical Temperature Scale of 1968. Temperatures are given to greater precision than the experimental uncertainties because they are used as calibration points of the International Practical Temperature Scale (see section 1.8).

	T/K	t/°C	*Estimated uncertainty*[3]/K
Standard			
Triple point of water[1]	273.16	0.01	exact by definition
Primary			
Triple point of hydrogen[2]	13.81	−259.34	0.01
Boiling point of hydrogen[2] at 25/76 atm pressure	17.042	−256.108	0.01
Boiling point of hydrogen[2] at 1 atm pressure	20.28	−252.87	0.01
Boiling point of neon at 1 atm pressure	27.102	−246.048	0.01
Triple point of oxygen	54.361	−218.789	0.01
Boiling point of oxygen at 1 atm pressure	90.188	−182.962	0.01
Boiling point of water at 1 atm pressure	373.15	100.00	0.005
Melting point of zinc at 1 atm pressure	692.73	419.58	0.03
Melting point of silver at 1 atm pressure	1235.08	961.93	0.2
Melting point of gold at 1 atm pressure	1337.58	1064.43	0.2

[1] The water should have the isotopic composition of ocean water.
[2] There are two types of molecular hydrogen, *ortho* and *para*. The reference points require the two types to be in their equilibrium proportions at the temperatures concerned.
[3] But see section 1.6 for the implications of more recent measurements.

accurate work perhaps 10^{-3} m^3). The volume is kept constant by adjusting the height of the right hand arm of the manometer so that the level of the mercury in the left stays constant. Corrections have to be made for the 'dead space' (the volume of gas not at the temperature being measured), changes of atmospheric pressure, variation of the density of mercury with temperature, expansion of the bulb, and so on. By using a gas which only liquefies at low temperatures (helium or hydrogen, for example) deviations from the ideal gas law are negligible unless the thermometer is being used for low temperature measurements, in which case the pressure must be kept low.

When the various corrections may be neglected, a constant volume gas thermometer may conveniently be calibrated at the ice point and the thermodynamic temperature found using the formula

$$T/\mathrm{K} = 273.15\, p/p_i \qquad (1.10)$$

Because they are inconvenient, gas thermometers are rarely used and, in order to be able to calibrate other thermometers accurately, the thermodynamic temperatures of several reference points have been determined with great precision using gas thermometry. Table 1.1 lists the temperatures of the standard and primary reference points as adopted in 1968.

Thermometers based on the *expansion of liquids* are very convenient for many purposes and, although they are not particularly accurate, they can be made very sensitive. Over a moderate range of temperature their scale is nearly linear in thermodynamic temperature (i.e. expansion is directly proportional to increase in thermodynamic temperature). Commonly used liquids are mercury and ethyl alcohol, which cover the ranges −39 to +350 °C and −117 to +78 °C respectively.

Thermometers based on the *expansion of solids* are relatively crude, but convenient for some applications such as the activation of thermostats or for driving the pens of thermographs. Usually, thin strips of dissimilar metals are bonded together to form a *bimetallic strip* which flexes as the temperature is changed, because the metal on one side expands more than the metal on the other (figure 1.15).

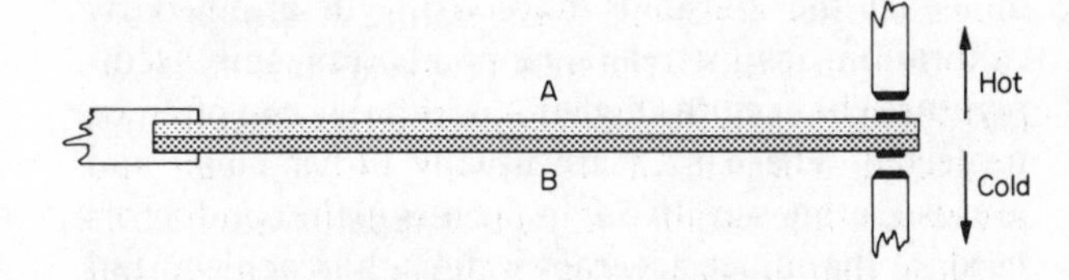

Fig. 1.15 A simple thermostat. If metal B has a greater expansivity than A, the bimetallic strip will flex upwards if the temperature rises and connection will be made with the upper contact.

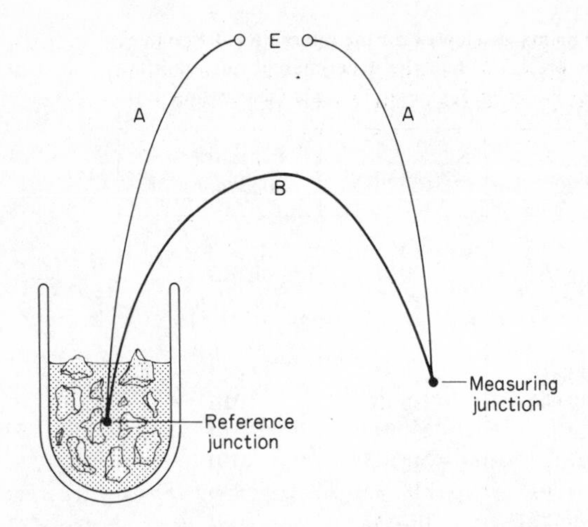

Figure 1.16 A thermocouple circuit using metals *A* and *B*. One junction is held at a fixed reference temperature (at the ice point, for example) and the other is at the temperature to be measured. If the thermoelectric effects in the two metals are unequal, there is a net e.m.f. *E* around the circuit when the junctions are at different temperatures.

Thermocouples make use of the variation with temperature of the *Seebeck e.m.f.*, the e.m.f. developed in a circuit of dissimilar conductors when the junctions are at different temperatures (figure 1.16). The e.m.f. arises because increasing the temperature of the conductor increases the thermal motions of the electrons which, in turn, makes them want to expand towards a cooler region. The extent to which they tend to move away from the hot regions depends on the conductor concerned so that, with a circuit of dissimilar conductors, the effects do not cancel one another and there is a net e.m.f. around the circuit. If one junction is kept at a constant temperature, the e.m.f., E, developed depends on the temperature t of the other junction in a non-linear way (i.e. it is not directly proportional); but the relationship can usually be represented reasonably accurately over a moderate range of temperature by a few terms of a power series:

$$E = a_0 + a_1 t + a_2 t^2 + a_3 t^3 + \dots \qquad (1.11)$$

in which the constants have to be determined by calibration against reference points. For small temperature changes, the higher power terms can often be neglected. The e.m.f.'s are usually rather small and are also rather sensitive to impurities in the conductors used, so that much accuracy is difficult to achieve, but thermocouples are often very convenient: they cover a large temperature range, they may be made very small so as to be able to explore regions where the

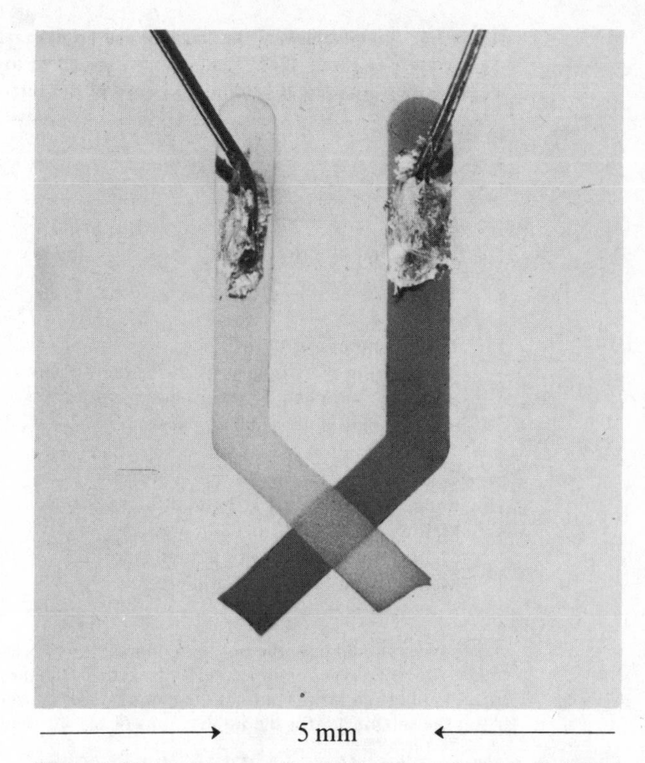

Figure 1.17 A thermocouple of thin films of copper and lead deposited on glass to measure the surface temperature. Connection to the films is made with thin wires of the same metals, soldered on with indium.

temperature is varying (figure 1.17), and they can respond rapidly. They are also useful for remote sensing (e.g. measuring the temperature in a furnace), since long connecting wires do not affect their operation. The approximate sensitivities (a_1 in equation *1.11*) of several commonly used thermocouples are listed in table 1.2.

Table 1.2 Approximate sensitivities of common thermocouples

Thermocouple	*Sensitivity/$\mu V\ K^{-1}$*
copper—constantan[a]	40
iron—constantan[a]	50
chromel[b]—alumel[c]	40
Pt—Pt + 10% Rh	6

[a] also called Eureka: 60% Cu, 40% Ni.
[b] 90% Ni, 10% Cr.
[c] 95% Ni plus Al, Si, Mn.

Resistance thermometry traditionally uses the variation with temperature of the electrical resistance of metals. The resistance increases with increasing temperature because the thermal motions (vibrations) of the atoms *scatter* (deflect) the electrons which are carrying the current and so impede their motion. As the temperature rises, the thermal motions increase, so the scattering increases and therefore the resistance also. Metal resistance thermometers again cover a very wide range of temperature and are useful from about 10 K to 1700 °C. For very accurate work, platinum is usually used because it can be obtained very pure which improves its performance at low temperatures, but for less critical work copper, being cheap and readily available, is a popular choice. The resistance is usually measured by passing a known current through the thermometer and measuring the potential drop across it. One must be careful not to use currents which are so large as to warm the thermometer above the temperature of its surroundings. Except at low temperatures, the metal resistance thermometer is not far from linear in thermodynamic temperature and its behaviour can usually be represented sufficiently accurately by the first few terms of a power series:

$$R = R_0(1 + a_1 T + a_2 T^2 + \ldots)$$

where $R_0, a_1, a_2, \ldots$ are constants which have to be determined by calibration. Figure 1.18 shows a copper resistance thermometer.

The variation of the electrical resistance of *semiconductors* (the kind of material used in making transistors) also provides the basis for sensitive thermometers. In semiconductors at absolute zero there would be no electrons free to move and carry current. All would be in *bound states* (ones in which they are immobile) and the semiconductor would be insulating. As the temperature rises, thermal motions excite electrons out of bound states into states where they are free to move and carry current. As a result, the electrical resistance of semiconductors, unlike that of metals, *decreases* as the temperature rises. Over certain ranges of temperature, the number of electrons excited into mobile states varies with temperature according to the Boltzmann factor (page 53), $\exp - (\varepsilon/kT)$, where ε is the energy by which they have to be excited.* The conductivity is roughly proportional to the number of free electrons, so that it varies in a similar way with temperature. Because the dependence involves an exponential of temperature, the resistance will vary very rapidly when $kT < \varepsilon$. Thermometers using semiconductors can therefore be extremely sensitive; changes of 1 mK are easy to measure at normal temperatures. Also, by preparing semiconductors with appropriate values of ε, semiconducting thermometers can be made to cover the range from below 1 K to above 300 °C. For use

* ε is the Greek letter *epsilon*.

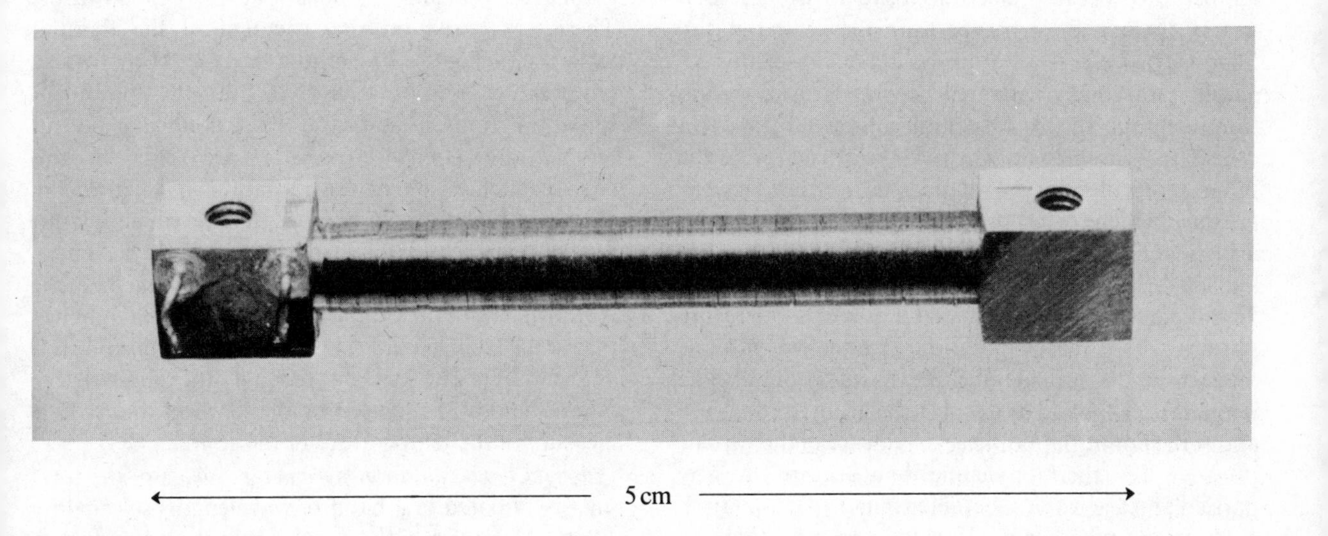

5 cm

Figure 1.18 A copper resistance thermometer used in low temperature research. Fine insulated copper wire is wound onto a copper former and varnished to it to ensure good thermal contact. The former is bolted tightly to the material whose temperature is to be measured.

around normal temperatures, *thermistors* are available commercially for a few pence in a variety of forms. They are small and robust. In some cars they are used as sensors for radiator temperature gauges. Figure 1.19 shows a typical thermistor characteristic.

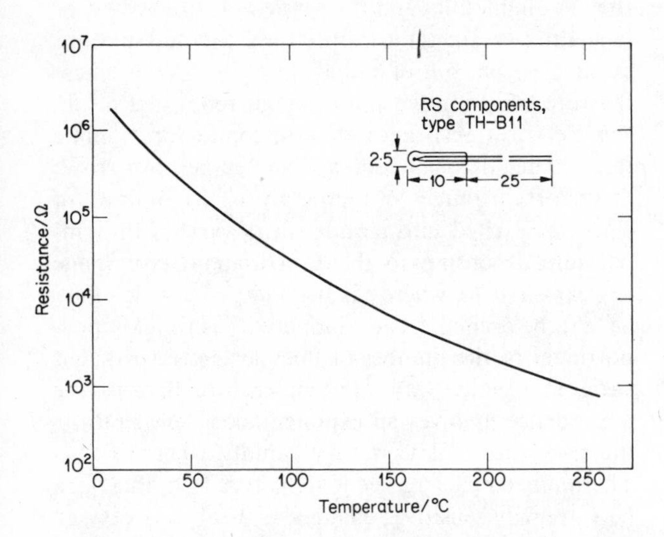

Figure 1.19 A typical thermistor characteristic. Note that the resistance scale is logarithmic. The inset shows the form of the thermistor with dimensions given in millimetres.

In semiconductors, the resistance varies with temperature because electrons have to use thermal energy to help them escape from the bound states. Such a process is said to be *thermally activated.* Another thermally activated process is evaporation from a liquid. In this case, molecules need the extra energy to help them escape from the attraction of the other molecules in the liquid. As a result, *vapour pressure,* which corresponds to how easily they escape, also depends strongly on temperature (section 5.6) and can be used as a means of measuring temperature. This method is particularly useful in low-temperature physics where the evaporation of liquefied gases is often used to provide the means of cooling. The temperature is found by simply measuring the pressure above the boiling liquid. The dependence of the vapour pressure on thermodynamic temperature in any particular case has to be determined in a separate calibration experiment. Vapour pressure thermometry is especially important from about 5 K to below 0.3 K where liquid helium is used as a refrigerant. Very accurate tables are available for both isotopes of helium relating vapour pressure to thermodynamic temperature. Figure 1.20 shows the vapour pressure of helium as a function of temperature. At normal temperatures, vapour pressure is sometimes used to actuate certain types of thermostat.

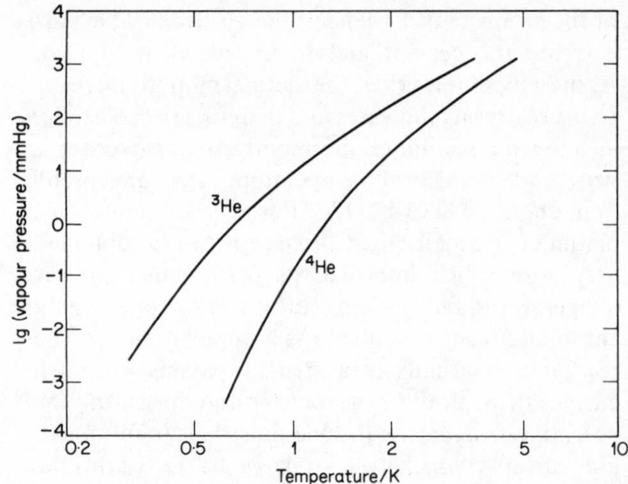

Figure 1.20 The temperature dependence of vapour pressure of liquid ^{4}He and liquid ^{3}He. ^{4}He is the common isotope. The proportion of ^{3}He in natural helium is only 1.3 ppm. Note that both scales are logarithmic.

None of the thermometers described above is useful far above 1500 °C. The range of *radiation pyrometers* extends from about 500 °C upwards. These are based on measurement of the thermal radiation (chapter 6) emitted by the body whose temperature is to be measured. Both the amount of radiation emitted and also its colour depend on temperature. *Optical pyrometers* are based on the change in intensity of the radiation at a particular wavelength and use an arrangement in which a lamp filament is viewed against the hot body as background (figure 1.21). The current through the filament is adjusted until the brightness of the filament matches that of the background. The filament then 'disappears' (figure 1.22). The current through the lamp determines the power radiated by the filament and so is a measure of the temperature of the source.

Broad band radiation pyrometers measure the total energy radiated in a band of wavelengths selected by filters. These are based on the Planck radiation law (page 101). There are also *total radiation pyrometers* which measure the total energy (of all wavelengths) radiated by the hot body, and these are based on

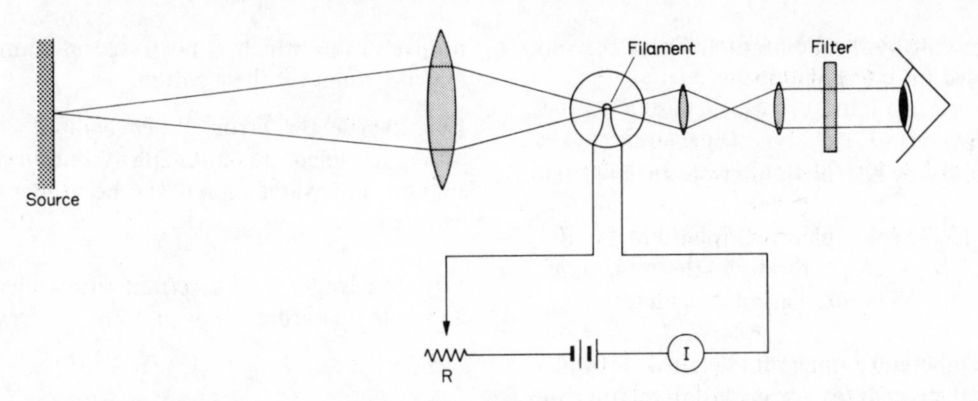

Figure 1.21 The elements of a disappearing filament optical pyrometer. The first lens forms an image of the source at the filament. The filament is viewed against the source as background through a filter. The current through the filament is adjusted by changing R until the source and filament are at the same temperature. The quality of radiation from each is then the same, and the filament disappears.

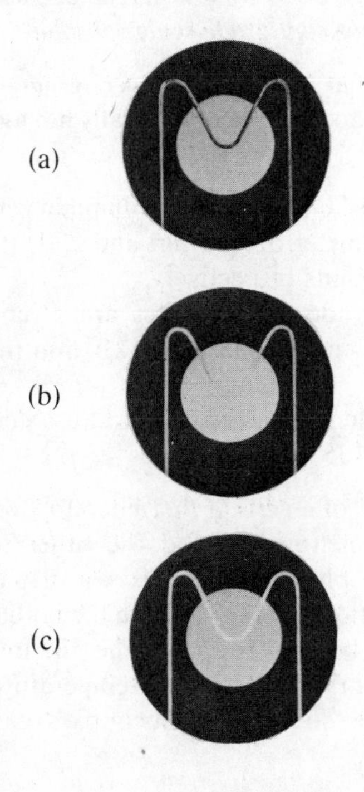

Figure 1.22 The filament of a 'disappearing filament' optical pyrometer. The filament is (a) colder than, (b) at the same temperature as, and (c) hotter than the source of radiation.
(Photograph by courtesy of Foster Cambridge Limited)

the Stefan–Boltzmann law (page 98) which relates total radiated power to temperature.

The useful ranges of common types of thermometer are given in table 1.3.

Table 1.3 Useful ranges of common thermometers

Type	*Range*/K	*Range*/°C
gas	3 – 1500	−270 – 1200
expansion of liquids	150 – 530	−120 – 260
thermocouples	3 – 2030	−270 – 1760
metal resistance	13 – 2000	−260 – 1750
semiconductor	1 – 600	−272 – 330
radiation pyrometers	above 750	above 480

1.8 THE INTERNATIONAL PRACTICAL TEMPERATURE SCALE

Because of the practical difficulties involved in precise measurement of thermodynamic temperature by gas thermometry, the Seventh General Conference of Weights and Measures in 1927 decided to establish a set of internationally agreed procedures which could be used for accurate temperature measurement in standards institutions and laboratories throughout the world. The scale is defined in terms of (a) a set of reference points whose thermodynamic temperatures have been determined with great accuracy, and (b) a set of recommended methods for interpolating between the reference points. The details are revised from time to time by the International Committee of Weights and Measures. The fixed points listed in table 1.1 are the ones used by the International Practical Temperature Scale of 1968, and the temperature values shown in the table are those allotted to the fixed points by the Committee. Three thermometers are used for interpolation:

Range	*Instrument*
13.81 K to 903.89 K	platinum resistance thermometer
903.89 K to 1337.58 K	platinum/(platinum + 10% rhodium) thermocouple
above 1337.58 K	radiation pyrometer

For each temperature interval, detailed formulae and numerical procedures are given for relating the thermometer readings to the thermodynamic temperature. The International Practical Temperature Scale is therefore essentially an *empirical* scale whose fixed points and interpolation procedures are chosen so that the resulting number is as close as possible to the true thermodynamic temperature. Full details of the scale and much useful experimental information are given in the booklet, *The International Temperature Scale of 1968* (Her Majesty's Stationery Office).

PROBLEMS

Laws of physics; macroscopic and microscopic approaches

1.1 Explain what is meant by a scientific 'law'. At what levels may such laws be formulated?

Systems, states, system variables, degrees of freedom, constraints. Work-like and thermal interactions; thermal contact; thermal isolation; adiabatic changes.

1.2 Explain the terms in italics. When two *systems* are placed in *thermal contact*, we usually find that they do not remain in their original *states*.

1.3 Explain what is meant by a *constraint*. Give an example of (a) a mechanical and (b) a thermal constraint.

1.4 Distinguish between *work-like* and *thermal* interactions. Give three examples of each kind which may be found in the home.

Thermal equilibrium and temperature. Zeroth law. Isotherms.

1.5 Explain how it follows from the zeroth law of thermodynamics that temperature is a universal property: one which is possessed in common by all systems whatever their nature.

1.6 Expose the error: When sodium is placed in water, a violent reaction always occurs; therefore sodium and water can never be at the same temperature.

1.7 The length L of a certain wire under tension F and at temperature t is given by

$$L = L_0(1 + \alpha t + \beta F)$$

where L_0, α and β are constants. Draw a sketch graph of L against F showing a series of isotherms.

Thermometric quantities, empirical scales of temperature. Centigrade scales; their differences. Gas thermometry; perfect gas centigrade scale; absolute zero.

1.8 Explain what is meant by a *centigrade temperature scale.* Why do such scales generally not agree with one another?

1.9 The lengths of the mercury column in a mercury-in-glass thermometer are 30 mm and 290 mm at the ice and steam points respectively.
a) What centigrade temperatures are given by the thermometer when the column is (i) 210 mm, (ii) 25 mm long?
b) What is the length of the column for a centigrade temperature of 135°?

1.10 The e.m.f. of a certain thermocouple when one junction is in melting ice and the other in water boiling at atmospheric pressure is +4.10 mV. When the second junction is removed from the boiling water and placed in boiling propane, the thermoelectric e.m.f. is −1.60 mV. What is the temperature of the boiling propane on the thermocouple's centigrade scale?

1.11 A copper resistance thermometer is calibrated at the ice and steam points. The values of resistance were 101.4 Ω and 144.7 Ω respectively.
a) What centigrade temperatures would be indicated by (i) 513 Ω, (ii) 2.0 Ω?
b) What resistances would correspond to centigrade temperatures of (i) 60°, (ii) −230°?

1.12 The resistance of nickel varies with Celsius temperature approximately as

$$R = R_0(1 + b_1 t + b_2 t^2)$$

constants. For a certain nickel resistance thermometer, the values of the constants were $R_0 = 43.0\,\Omega$, $b_1 = 5.30 \times 10^{-3}\,\mathrm{K}^{-1}$, $b_2 = 6.51 \times 10^{-6}\,\mathrm{K}^{-2}$. Calculate the values of R for (a) 100 °C and (b) 150 °C. (c) What is 150 °C on the thermometer's centigrade scale?

Thermodynamic temperature; the kelvin

1.13 Explain why the unit of thermodynamic temperature is defined in terms of a single fixed point. Why are gas thermometers important in connection with determinations of thermodynamic temperatures?

Celsius temperature

1.14 Explain the difference between the Celsius and perfect gas centigrade temperature scales. Which of the following statements is correct? Give your reasoning in each case.
a) The temperature of the normal boiling point of water is exactly 100 °C.
b) The temperature of the triple point of water is exactly 273.16 K.
c) Absolute zero is exactly −273.15 °C.

Common thermometers: gas, expansion of liquids and solids, thermocouples, resistance of metals and semiconductors; vapour pressure thermometry; radiation pyrometers

1.15 What factors should be taken into account when choosing a thermometer for a particular application? What types of thermometer might you use for measuring
a) the melting point of paraffin wax;
b) the temperature of a silver crystal in order to control the furnace in which it is being heated, so as to keep its temperature about 20 K below the melting point (962 °C);
c) the variation of body temperature during sleep;
d) the boiling point of neon (24.48 K);
e) the temperature difference across the glass of a window;
f) the melting point of platinum (1772 °C);
g) small changes in the temperature of nitrogen boiling at about 68 K;
h) the temperature 300 m below the surface of the sea;
i) the temperature at the point of impact when a steel ball rebounds from a copper plate.

1.16 A constant volume air thermometer is used on a day when the atmospheric pressure is 750 mmHg. When calibrated at the ice point, the difference h in the levels of the mercury in the manometer (figure 1.14) is +15 mm. Assuming that the behaviour of the air is close to that of an ideal gas,
a) what temperature is indicated by a difference of +160 mm?
b) what temperature is indicated by a difference of −25 mm?
c) what is the difference of levels for a temperature of 120 °C?
d) would it be necessary to recalibrate the thermometer for use on another day when the atmospheric pressure is 770 mmHg?

1.17 A chromel-alumel thermocouple has a mean sensitivity between 0 °C and 100 °C of 41 $\mu\mathrm{V\,K}^{-1}$. If the thermoelectric e.m.f. is measured with a potentiometer which can be set and read to an accuracy of 1.0 μV, what will be the percentage uncertainty when the thermocouple is used to measure a temperature difference of about 10 °C?

1.18 A copper-constantan thermocouple is to be used with the reference junction in ice to measure temperatures in the range 0 °C to 300 °C. It is calibrated at the ice and steam points, and at the melting point of tin. The following values were found for the e.m.f. E with the measuring junction at calibration temperatures t:

t/°C:	0	100	232
E/mV:	0	4.28	11.02

a) What is the temperature of the tin point on the thermocouple's centigrade scale?

If the temperature dependence of the e.m.f. is to be represented by the formula

$$E = a_0 + a_1 t + a_2 t^2$$

b) substitute in each set of values and solve the equations for a_0, a_1 and a_2.
c) Find the value given by the equation for E at 183 °C, the melting point of 60 Sn:40 Pb solder.
d) If the measured value of the e.m.f. at the melting point of the solder is 8.39 mV, estimate the accuracy with which temperatures will be determined in the 0–300 °C range using your values of a_0, a_1 and a_2 in the quadratic formula to relate E and t.
e) What is the most important factor contributing to the uncertainty?

1.19 When using a resistance thermometer, it is sometimes sufficiently accurate to take the resistance as varying linearly with thermodynamic temperature instead of using a more accurate formula to relate resistance to temperature. In this question you examine the errors involved.

A platinum resistance thermometer is calibrated at the ice and steam points. The values of resistance found were $R_i = 105.2\ \Omega$ and $R_s = 146.4\ \Omega$.

a) What would be the value of the resistance at 50 °C on the (linear) platinum centigrade scale?

The temperature variation of the resistance of platinum is better represented by the quadratic formula

$$R = R_0(1 + at + bt^2)$$

where R_0 is a constant, t is (Celsius) temperature and a and b have the values $a = 3.98 \times 10^{-3}\ \mathrm{K}^{-1}$ and $b = -6.04 \times 10^{-7}\ \mathrm{K}^{-2}$.

b) What would the value of R be at 50 °C?

c) If, when using the thermometer, instead of using the more accurate quadratic formula to find t from the measured resistance, the Celsius temperature is taken to be the same as the centigrade temperature, is the supposed Celsius temperature too high or too low?

d) Roughly what error is involved near 50 °C in taking the resistance to vary linearly with temperature?

e) In view of the accuracy of the measurements, is it worth using the more accurate formula?

1.20 The thermistor characteristic shown in figure 1.19 has a logarithmic vertical scale so that equal vertical intervals correspond to equal *fractional* changes in resistance.

a) By measuring from the figure, find the values of a and b in the equation $\lg R = a + bt$ which give the best fit between the equation (which would plot as a straight line on the figure) and the curve between 50 °C and 100 °C.

b) Hence, if the resistance can be measured to 1%, estimate the minimum *change* in temperature which can be detected in this region.

[You may wish to use the fact that $\ln x \approx 2.3 \lg x$.]

The International Practical Temperature Scale

1.21 Explain what is meant by the *International Practical Temperature Scale*. Why is it needed?

2 Internal Energy, Heat and Work

2.1 THE FIRST LAW OF THERMODYNAMICS

The first law of thermodynamics is about the relationship of heat, work and energy. Perhaps it seems fairly obvious to us that when a system is in a given state so that the system variables p, V, T, etc. have a definite set of values, then the system also has a definite total energy. But this was not obvious to the scientists of the last century, nor, in fact, is it as self evident as we might at first think. The macroscopic state of a given mass of gas, for example, is fixed once we set the values of any two of the variables p, V, T. Having set the two, we shall always get the same results for other measurements we might make on the gas—that is what being in a definite state means. But we know that within the gas the molecules are colliding all the time with each other and with the walls of the container, and we know that in these collisions energy is exchanged. It is not obvious that whenever we observe this gas with given values for, say, pressure and volume, its total energy is always the same, for there are so many ways in which energy could be shared among the molecules. The first purpose of the first law of thermodynamics is to assert that a given state always has the same total energy associated with it. The total energy is called the *internal energy* and is given the symbol U. Its normal unit is the joule.

Any quantity which always takes a definite value for each state of a system is called a *function of state*. Thus, the first law asserts that internal energy is a function of state: for each set of values of p, V, T, etc., the energy U has a definite fixed value.

The basis for this assertion lies in experiments in which a given change of state is produced in a thermally isolated system by doing work on it in a variety of ways: it is found that the total amount of work required for the given change is always the same regardless of how the work is done. This was the famous work of James Prescott Joule (1818–1889). The *principle* of his experiments is illustrated in figure 2.1. (This is not, in fact, how Joule did his experiments.) The system is a mass of gas contained in a cylinder fitted with a piston and thermally isolated from its surroundings. Also in the cylinder, and part of the system, is an electrical heater. As seen from outside, work may be done on the system in two ways: mechanically by pushing the piston in, and electrically by applying a voltage to the terminals of the heater. Possible paths between two states P and Q are illustrated in figure 2.2. The path via A consists of an

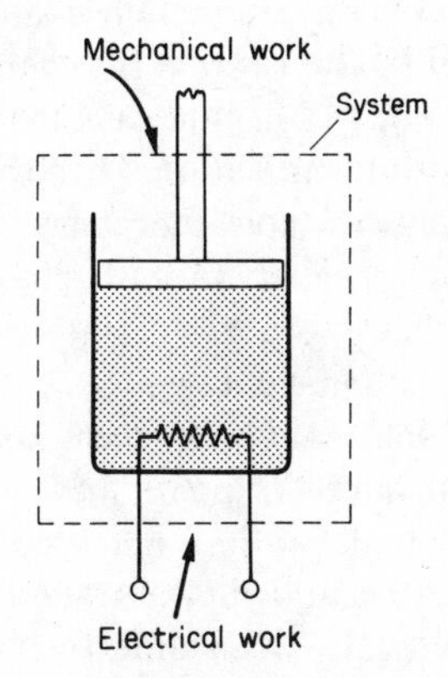

Figure 2.1 A system with which we may illustrate the first law of thermodynamics

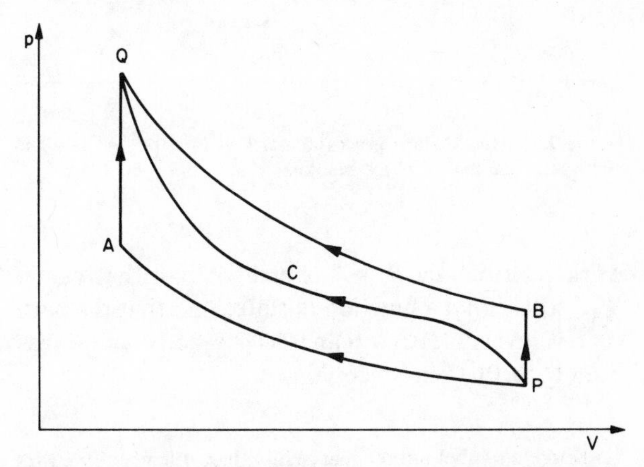

Figure 2.2 Three possible paths between the states *P* and *Q* of the system of figure 2.1

(adiabatic) compression followed by electrical work (which is *internally* converted to heat). The path via B reverses the order; and to follow the path marked C, compression and electrical work have to be carried out simultaneously at carefully adjusted rates. The result of such experiments on thermally isolated systems is that *the total work required is always the same regardless of how the change is brought about.* Since doing work on the system changes its energy, this is equivalent to saying that the change* in internal energy, $\Delta U = U_Q - U_P$, is always the same however the system passes from P to Q. Then each state of the system must always have the same internal energy because, no matter what we do to the system, whenever we come back to a given state we must always come back to the same energy.

Note that if internal energy were *not* a function of state we would have the possibility of being able to devise a perpetual motion machine. Suppose, for example, that we could find a system that could be changed adiabatically from state P to state Q by two paths A and B (figure 2.3). Suppose then that the work required via B, W_B, is greater than that via A, W_A, (so U is *not* a function of state), and that the path via B is reversible (so we can trace it exactly backwards from Q to P), then we could go:

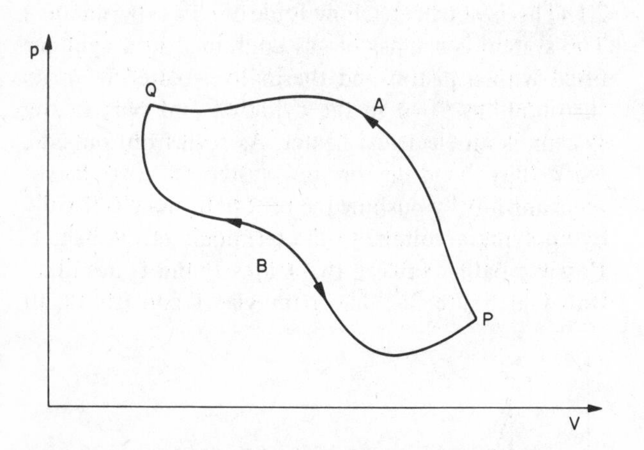

Figure 2.3 Hypothetical processes which violate the first law and allow perpetual motion of the first kind

a) from P to Q via A which costs us W_A,
b) back from Q to P via B during which we get W_B *out* of the system since we are tracing the path backwards.

In going around the loop $PAQBP$ once, we get *out* of the system

$$W_B - W_A > 0$$

We get net energy out. The cycle can be repeated indefinitely so the system amounts to a machine which manufactures an inexhaustible supply of energy. This is known as *perpetual motion of the first kind* (figure 2.4). (For perpetual motion of the second kind see page 104.)

The second function of the first law is to state that heat is a form of energy. The experiments on thermally isolated systems lead us to the conclusion that internal energy is a function of state. But if we examine changes *without* the condition of thermal isolation so that heat can flow, we find that the amount of work required for a given change is no longer constant. We therefore either have to abandon the idea that internal energy is a function of state, or we have to conclude that heat is a form of energy so that both work and heat contribute to the total energy change of the system.

We may therefore summarize the *first law of thermodynamics* by the two statements:

Internal energy is a function of state; heat is a form of energy.

Together, these amount to a restatement of the principle of conservation of energy in a form appropriate for discussing thermal processes. Applied to changes of state, they are expressed algebraically by the equation

$$\Delta U = W + Q \qquad (2.1)$$

where

ΔU is the change in internal energy,
W is the work done *on* the system,
Q is the heat which flows *into* the system.

The normal unit for all terms in the equation is the joule. It is important to note the sign convention. For the terms on the right hand side, the sign is taken as positive when energy is added to the system. If heat flows from system to surroundings, Q is negative. Similarly, if the system does work on the surroundings, W is negative.

* Δ before a symbol means 'change in'. Thus, if temperature rises from 15 to 20 °C, $\Delta T = 5$ K. Δ is the capital form of the Greek letter *delta*.

Figure 2.4 Maurits Escher's *Waterfall* shows perpetual motion of the first kind. (Photograph by courtesy of The Escher Foundation, Haags Gemeentemuseum, The Hague.)

It is also important to notice a difference between the terms on the two sides of equation *2.1*. In a given change of states, *however the change is made*, ΔU always has the same value because U is a function of state so that $\Delta U = U_{\text{final}} - U_{\text{initial}}$ is fixed. But we only know the values of W and Q if we know *how* the change takes place. If, in the experiment illustrated in figure 2.1, we had counted the heater as being *outside* the system so that the electrical work was converted to heat before entering (figure 2.5), then we would have taken the electrical work as Q and the compression as W. By varying the path we now have some freedom of choice as to how $U_Q - U_P$, which is fixed, is divided between heat and work. Thus, *heat and work are not functions of state*: there is no sense in which one can say that a system contains so much heat and so much work. The division of energy changes into two components (heat and work) concerns simply the form in which the energy enters the system: we call the transfer of energy 'work' if it involves an ordered process like a force pushing a piston; we call it 'heat' if it involves energy in a disordered form like the thermal motions of atoms in a solid or of molecules in a gas.

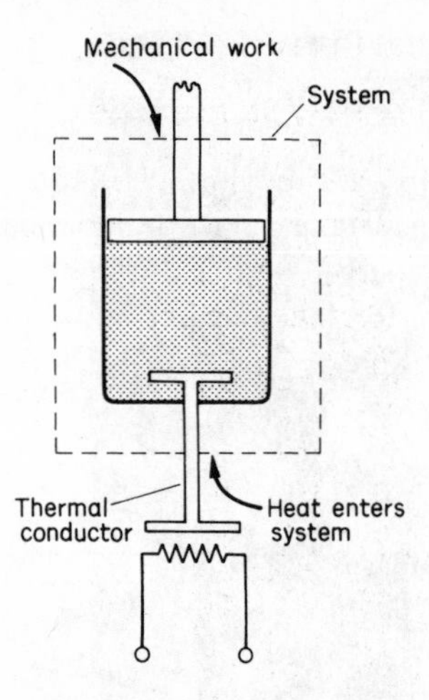

Figure 2.5 A modified version of the apparatus of figure 2.1 in which the electrical work is converted to heat before entering the system

We note that the heater in figures 2.1 and 2.5 converts work into heat. The action of the potential difference driving the current through the heater is work; but the energy leaves the heater as heat. We sometimes say that energy is *degraded* when it passes from an ordered form (work) to a disordered form (heat). If the heater were kept at constant temperature, it would be unchanged after electrical work had been done on its terminals and we would conclude that *all* the electrical work must leave as heat. In fact, we can always arrange for *complete* conversion of work into heat; friction is a mechanical means. But it turns out that the reverse is not possible: heat cannot be fully converted into work. This is the sense in which energy becomes degraded on conversion to heat; it can never fully be recovered in ordered form. This irreversibility in nature is the subject of the *second law of thermodynamics* which we shall discuss very briefly in chapter 7.

We can use equation *2.1* to help us give a formal definition of the terms *hotter* and *colder*. If we put into thermal contact two bodies which are not at the same temperature, heat will flow from one to the other until thermal equilibrium is established. If the total heat exchanged is Q we will have $\Delta U = -Q$ for one body and $\Delta U = +Q$ for the other (figure 2.6). We say that the body for which ΔU was negative (heat flowed out) was the hotter, and the body with ΔU positive (heat flowed in) the colder. Hotter and colder are thus comparative terms concerning the direction of heat flow when systems are put into thermal contact.

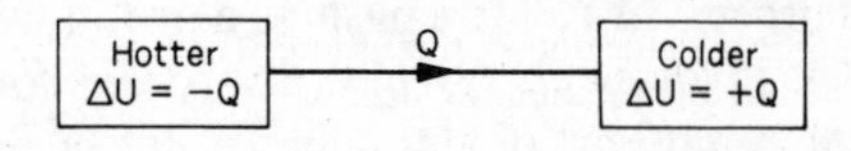

Figure 2.6 Heat flows from a hotter body to a colder

2.2 SOME FORMS OF WORK

The work done when a *force* **F** moves its point of application an infinitesimal distance dx is*

$$\mathrm{d}W = F_x\,\mathrm{d}x \qquad (2.2)$$

where F_x is the component of the force in the direction of the displacement.† To calculate the work for a finite displacement, the law connecting **F** to **x** must be known so that the expression for the differential quantity dW can be integrated. For example, a spring which obeys Hooke's law has

$$F_x = \alpha x$$

where α is the spring constant and x the extension.‡ A spring is little affected by temperature or pressure, so there is normally only one significant degree of freedom, namely the extension. Then

$$\mathrm{d}W = \alpha x\,\mathrm{d}x$$

and, the work done in increasing the extension from x_1 to x_2 is

$$W = \int_{x_1}^{x_2} \mathrm{d}W = \int_{x_1}^{x_2} \alpha x\,\mathrm{d}x = \tfrac{1}{2}\alpha(x_2^2 - x_1^2) \qquad (2.3)$$

Alternatively, since this is a case where force is linearly proportional to displacement, we may use the argument that the work done is (average force) × (displacement):

average force $= \tfrac{1}{2}\alpha(x_2 + x_1)$

displacement $= (x_2 - x_1)$

so work $= \tfrac{1}{2}\alpha(x_2 + x_1)(x_2 - x_1) = \tfrac{1}{2}\alpha(x_2^2 - x_1^2)$

* d placed before a variable is the differential notation of calculus and means 'an infinitesimally small change in'. Δ, which we introduced on page 18 is used for finite changes.
† Force and displacement are vector quantities, which are indicated by heavy type. It is only the component of a force which is parallel to a displacement which does work, the magnitude of the component being $F\cos\theta$, where θ is the angle between **F** and d**x**. The vector notation for multiplication including the cosine of the angle is the *dot product*:

$$\mathrm{d}W = \mathbf{F}\cdot\mathrm{d}\mathbf{x} = F\cos\theta\,\mathrm{d}x = F_x\,\mathrm{d}x$$

‡ α is the Greek letter *alpha*.

This argument is only correct when one quantity is linearly proportional to the other, and is only given here for those who have no calculus. Those who can integrate should *always* carry the calculation through properly in order to avoid loose thinking and mistakes in other cases.

The expression for the work done by a force can be used to derive that for work done by *hydrostatic pressure*. Consider a fluid in a cylinder with a frictionless piston of area A (figure 2.7). If the pressure in the fluid is p, the force necessary to hold the piston in equilibrium is pA. If the piston is now moved in a small distance $\mathrm{d}x$, the work done is

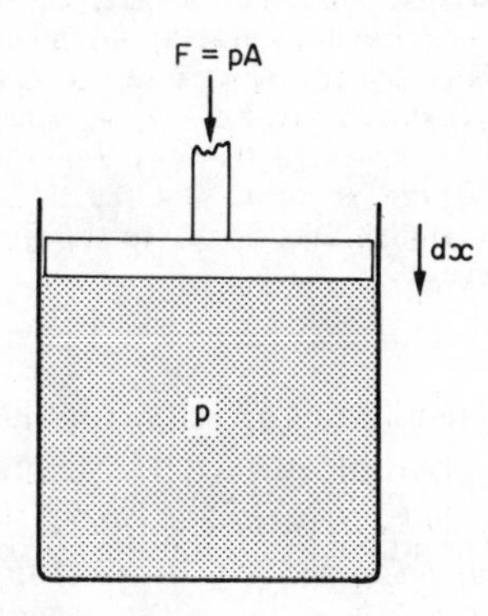

Figure 2.7 Calculation of work done by hydrostatic pressure

$$\mathrm{d}W = F_x\,\mathrm{d}x = pA\,\mathrm{d}x = -p\,\mathrm{d}V \qquad (2.4)$$

The negative sign is present because, when the piston moves in ($\mathrm{d}x$ positive), the volume decreases ($\mathrm{d}V$ negative), so $\mathrm{d}V = -A\,\mathrm{d}x$.

Again, to evaluate *2.4* for a finite change in any particular case, we need to know *how* the change takes place so that we may substitute for p in terms of V and then integrate. As examples we will take isothermal and adiabatic changes in an ideal gas. For an isothermal change, we use the ideal gas law (page 40),

$$pV = nRT$$

where n is the number of moles. This gives

$$\mathrm{d}W = -nRT\frac{\mathrm{d}V}{V}$$

which integrates to give

$$W = nRT\ln(V_1/V_2) \qquad (2.5)$$

where V_1 and V_2 are the initial and final volumes.*

If the change takes place adiabatically, p and V are related by $pV^\gamma = \text{constant}$ (page 42).† In this case, using K for the constant,

$$\mathrm{d}W = -K\frac{\mathrm{d}V}{V^\gamma}$$

Integrating,

$$W = -K\int_{V_1}^{V_2}\frac{\mathrm{d}V}{V^\gamma}$$

$$= \frac{K}{(\gamma-1)}[V^{1-\gamma}]_{V_1}^{V_2}$$

$$= \frac{K}{(\gamma-1)}(V_2^{1-\gamma} - V_1^{1-\gamma})$$

Substituting with

$$K = p_1V_1^\gamma = p_2V_2^\gamma$$

$$W = (p_2V_2 - p_1V_1)/(\gamma-1). \qquad (2.6)$$

In the case of *electrical work*, we start with the definition of the volt, according to which the rate of doing work (power) when a current I flows through a potential difference V is‡

$$\dot{W} = \frac{\mathrm{d}W}{\mathrm{d}t} = IV = V\dot{Q} = V\frac{\mathrm{d}Q}{\mathrm{d}t}$$

Here, Q is charge, not heat!

From the second and last expressions in this series of equalities,

$$\mathrm{d}W = V\,dQ \qquad (2.7)$$

This is identical to the electrostatic definition of potential:

work = (charge) × (potential difference the charge is moved through)

* ln is the standard notation for logarithm to the base e. The ln V appears through the integration of V^{-1}. The differential relationship is

$$\frac{\mathrm{d}}{\mathrm{d}x}(\ln x) = \frac{1}{x}$$

† γ is the Greek letter *gamma*.

‡ The notation with the dot, $\dot{W}$, means rate of change with respect to time. In the normal notation of calculus,

$$\dot{W} = \frac{\mathrm{d}W}{\mathrm{d}t}$$

When current is driven through a resistor of resistance R, we may substitute with Ohm's law $V = IR$ to find the rate at which the electrical work is dissipated as heat:

$$\dot{W} = I^2 R \tag{2.8}$$

Two electrical components which do not dissipate electrical work but *store* the energy are capacitors and inductors. In the case of a capacitor of capacitance C,

$$V = Q/C$$

$$dW = (Q/C)\,dQ$$

and integrating,

$$W = (Q_2^2 - Q_1^2)/2C \tag{2.9}$$

where Q_1 and Q_2 are the initial and final charges on the capacitor. This is analogous to the stretched spring formula (equation *2.3*). In both cases the energy is stored in potential form.

For an inductor of inductance L carrying a current I, we have, by definition of inductance,

$$V = L\dot{I}$$

so,

$$\dot{W} = LI\dot{I}$$

$$dW = LI\,dI$$

(cancelling the dt from both sides) and integrating

$$W = L(I_2^2 - I_1^2)/2 \tag{2.10}$$

where I_1 and I_2 are the initial and final currents. In this case, the energy is associated with motion of charge and is analogous to mechanical kinetic energy.

Surface tension is *defined* as the work required to create unit area of surface. Thus,

$$dW = \gamma\,dA \tag{2.11}$$

where γ is surface tension, and A area.
If we imagine the area of surface to be increased by means of some simple mechanical arrangement like that illustrated in figure 2.8, we see that γ is also the force per unit length exerted by the surface normal to any line in the surface. (This is a two-dimensional analogue of pressure.) Experimentally it is found that γ is independent of A and varies with T only. Then, for an isothermal stretching of a surface

$$W = \gamma(A_2 - A_1) \tag{2.12}$$

where A_1 and A_2 are the initial and final areas.

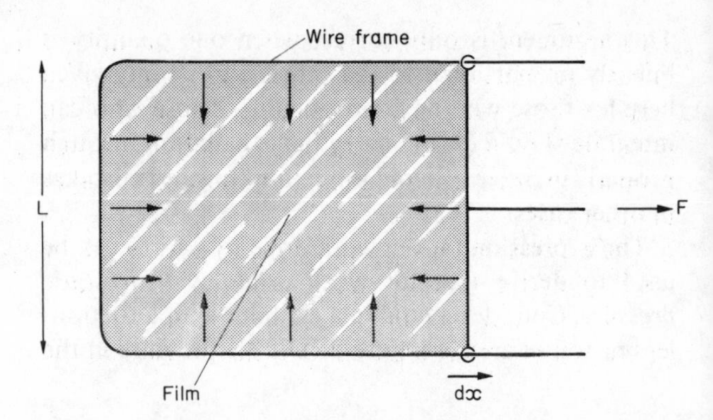

Figure 2.8 Relating work done in increasing the area of a film to the force per unit length exerted by the film. The film has two surfaces, so when the sliding wire moves forward a distance dx the area of new surface created is $2L\,dx$ and the work required $2\gamma L\,dx$. This is supplied by the force F moving through dx; so $2\gamma L\,dx = F\,dx$, and $F = 2\gamma L$. But the total length of surface pulling on the sliding wire is $2L$; so the force exerted by unit length of surface is γ.

We may use the expression for the work done in increasing area of surface to deduce the *pressure difference across a curved surface.* We consider the spherical case. Consider a drop of liquid of radius r suspended from the end of a fine capillary, the other end of which connects to a cylinder with piston which is also full of the liquid (figure 2.9). We ignore

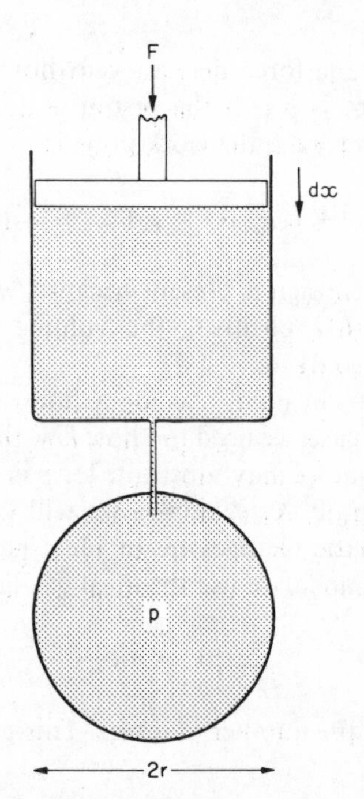

Figure 2.9 Calculation of the pressure difference across a curved surface using an argument based on work

gravitational forces which are irrelevant to the energy changes we are considering, and also assume the liquid incompressible. If there is a pressure difference p across the curved surface, a force F has to be applied to the piston to keep it in equilibrium. Now push the piston in a small distance dx. The liquid will be forced down the capillary and the drop will grow. The work required can only go into increasing the surface area A of the drop, so

$$dW = F\,dx = -p\,dV_{\text{cylinder}} = +p\,dV_{\text{drop}} = \gamma\,dA$$

But

$$V_{\text{drop}} = \tfrac{4}{3}\pi r^3$$

so

$$dV_{\text{drop}} = 4\pi r^2\,dr$$

and

$$A = 4\pi r^2$$

so

$$dA = 8\pi r\,dr$$

Substituting for dV_{drop} and dA in the last two expressions for dW we obtain

$$p = 2\gamma/r \qquad (2.13)$$

The same result may be obtained by simply considering the statics of a drop in terms of the tensional force in the surface. Consider a hemispherical drop of radius r sitting on a solid surface (figure 2.10). The total up-

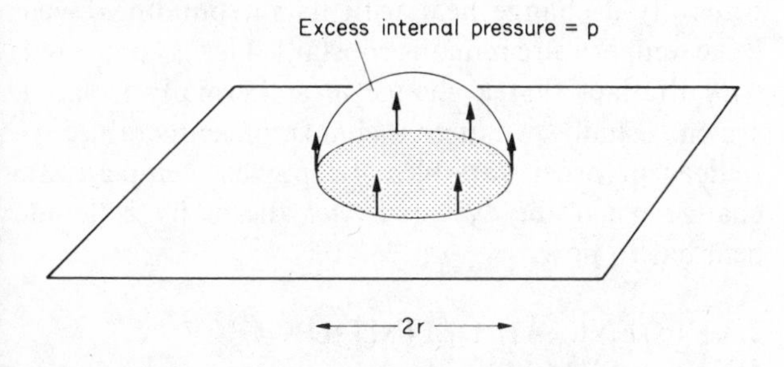

Figure 2.10 Calculation of the pressure difference across a curved surface using an argument based on static equilibrium

ward force exerted by the liquid surface around the perimeter of the base of the drop is $\gamma 2\pi r$. The whole system must be in mechanical equilibrium, so this upward force must be balanced by the effect of the excess pressure p inside the drop acting downwards on the area of flat surface within the perimeter: $p\pi r^2$. Thus

$$2\pi\gamma r = p\pi r^2$$

which again gives *2.13*.

We shall discuss the reason for the existence of surface tension in section 5.5.

2.3 HEAT CAPACITIES

If we add heat to a system its temperature will rise. Some systems will need a lot of heat to make their temperature rise by a given amount, others will need less. Heat capacities are a measure of the amount of heat required.

The general definition of heat capacity needs the formalism of calculus, and we will give that later. However, to introduce the idea, we will see what heat capacity means in a special but common case: very often, the amount of heat required to raise the temperature of a body by a certain number of kelvins does not depend significantly on temperature. In this case, the heat capacity may be taken as a constant. Then a simple starting definition would be

Heat capacity = heat absorbed per kelvin rise in temperature

or, in symbols,

$$Q = C\,\Delta T$$

where

Q is heat absorbed
C is heat capacity
ΔT is change of temperature.

The unit for C will be joules per kelvin: $\mathrm{J\,K^{-1}}$.

However, this definition is not quite complete because we must remember that simple systems have two degrees of freedom so that we do not know the way in which the system will respond to the change of temperature ΔT unless further information is given to tell us *how* the change takes place. The path of the change will be fixed if we apply a constraint. In the simplest cases, this is done by keeping some other variable constant. The variable kept constant is then added as a suffix. For example,

$$C_p = Q/\Delta T \qquad (2.14)$$

is the heat capacity at constant pressure. Q is the heat

absorbed as the temperature rises by ΔT, *pressure being kept constant.*

Similarly

$$C_V = Q/\Delta T \qquad (2.15)$$

is the heat capacity at constant volume, Q being the heat needed to raise the temperature by ΔT when the system is *prevented from changing its volume.* C_p and C_V are known as *principal heat capacities* because they are the most important in everyday physics.

Why are C_p and C_V different? Unless we prevent it from doing so, a system will normally expand as its temperature rises. If it is in surroundings which exert a pressure on it, the system will have to do work on the surroundings as it expands. In a normal laboratory experiment, for example, when an object expands, work has to be done in pushing back the atmosphere to make room for the increased volume the object takes up. Thus, when a system is heated at constant pressure, energy is required both to warm the system and to push back the surroundings. If, instead of allowing the system to expand, the volume is kept constant, no work is done on the surroundings, and the heat only has to supply the energy required to increase the thermal motions within the system (i.e. increase its temperature).

We may see more clearly what is involved here by applying the first law as it is expressed by equation *2.1*. We have

$$Q = \Delta U - W$$

where

Q is heat absorbed
ΔU is change of internal energy of the system
W is work *by* the surroundings *on* the system.

If the surroundings exert a constant pressure p_0,

$$W = -p_0 \Delta V$$

where ΔV is the change of volume of the system. Then, in general

$$Q = \Delta U + p_0 \Delta V$$

In the case of the heat capacity at constant volume, $\Delta V = 0$ and we have

$$C_V = \frac{Q}{\Delta T} = \frac{\Delta U}{\Delta T} \qquad (2.16)$$

while at constant pressure we have

$$C_p = \frac{Q}{\Delta T} = \frac{\Delta U}{\Delta T} + p_0 \frac{\Delta V}{\Delta T} \qquad (2.17)$$

$p_0(\Delta V/\Delta T)$ is the extra energy required to push back the surroundings. $\Delta V/\Delta T$ is the increase in volume per kelvin temperature rise. If this is small, C_p and C_V will be very nearly equal. This is why one often finds no distinction made between C_p and C_V for solids in tables of physical constants: solids expand so little that the principal heat capacities are essentially equal.

Since the amount of heat absorbed by a system in a given change of temperature depends on the mass of the system, it is sometimes useful to be able to refer to the heat capacity for a given mass. Thus we have (for the constant pressure case)

molar heat capacity,	C_{mp},	unit $\mathrm{J\,K^{-1}\,mol^{-1}}$
specific heat capacity,	c_p,	unit $\mathrm{J\,K^{-1}\,kg^{-1}}$

For the molar quantity we add a suffix m; for the specific heat capacity we simply use a small letter. These conventions are used with other variables; for example, molar volume (volume per mole) V_m and specific volume (volume per kilogram) v. When there is no possibility of confusion, the suffixes are often omitted. 'Specific heat capacity' is sometimes (misleadingly) abbreviated to 'specific heat'.

Flow of heat is not only associated with change of temperature, as we have seen in the above discussion where heat also supplies the energy for mechanical work. When the state of a system is changed, it will generally exchange heat with its surroundings even if the temperature remains constant. This is connected with the fact that a change in a thermally isolated system usually produces a change in temperature as well as in other variables. To prevent temperature change when the system is *not* thermally isolated, heat has to flow.

2.3.1 GENERAL DEFINITION OF HEAT CAPACITY

Heat capacities are generally not constant but depend on temperature. The specific heat capacity of diamond, for example, rises from $500\,\mathrm{J\,K^{-1}\,kg^{-1}}$ at room temperature to $1800\,\mathrm{J\,K^{-1}\,kg^{-1}}$ by 600 °C. The reason for this is connected with quantum theory, and need not concern us here; but it is clear that we need a more general definition which can take account of this.

If a heat capacity is constant and we plot heat added against rise in temperature we will get a straight line since every increase of 1 K needs the same amount of heat (figure 2.11). The heat capacity is the gradient of

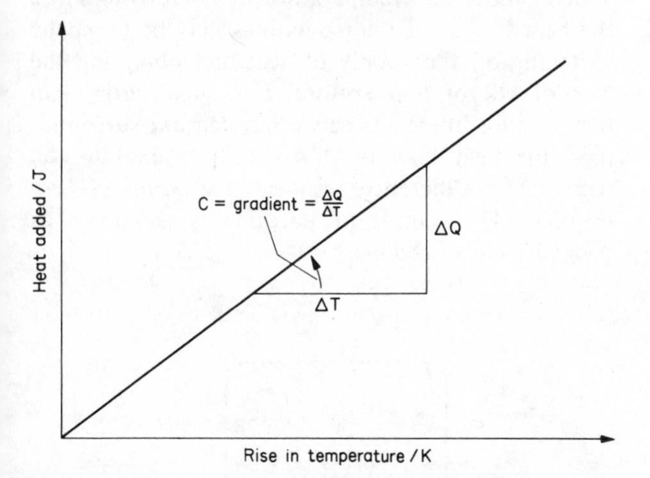

Figure 2.11 If a heat capacity is constant, the plot of heat added against temperature rise is a straight line of gradient *C*

this line. If the heat capacity is not constant, the graph of heat added against temperature will not be a straight line (figure 2.12). The obvious generalization

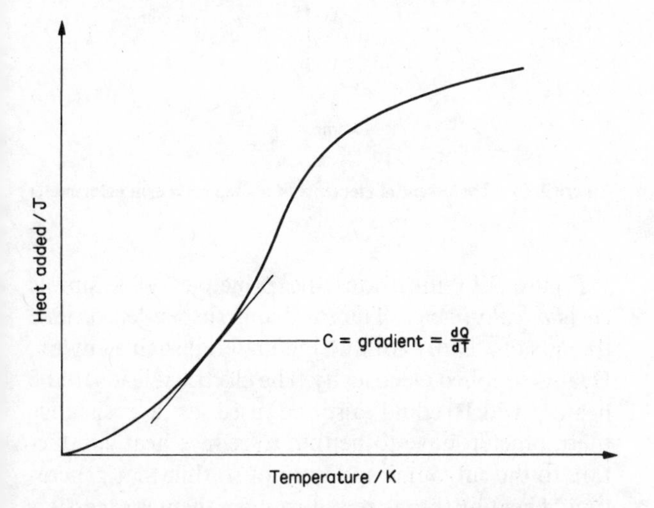

Figure 2.12 When a heat capacity is not constant, the plot of heat added against temperature is not a straight line. The heat capacity at any temperature is the gradient of the curve at that temperature

is to take the heat capacity at a given temperature as the gradient of this curve at that temperature. The equivalent of our original definition would then be

Heat capacity = gradient of graph of heat absorbed/temperature

which we might be tempted to write, in the usual notation of calculus

$$C = \frac{\mathrm{d}Q}{\mathrm{d}T}$$

where $\mathrm{d}Q$ is the infinitesimal amount of heat absorbed in the infinitesimal temperature rise $\mathrm{d}T$. But we must remember the need for a constraint if the heat absorbed in a given change of temperature is to be fixed. The correct notation is

$$C_p = \left(\frac{\partial Q}{\partial T}\right)_p \qquad (2.18)$$

which is the rate of absorption of heat with change of temperature (the gradient of the Q/T graph) when the pressure is kept constant. The right hand side of this equation is known as a *partial differential** which means that $\mathrm{d}Q$ depends on more than one variable. Note the use of the curly ∂ in a partial differential instead of the usual d.

When a heat capacity is not constant, the total heat absorbed in a finite change of temperature has to be found by integration. From *2.18*

$$\mathrm{d}Q = C_p\,\mathrm{d}T \qquad (p \text{ constant})$$

so

$$Q = \int_{T_1}^{T_2} C_p\,\mathrm{d}T$$

The integration can be performed when it is known how C_p depends on temperature.

In *differential form* the first law becomes

$$\mathrm{d}U = \mathrm{d}Q + \mathrm{d}W$$

* Formally, the partial differential is defined by

$$\left(\frac{\partial x}{\partial y}\right)_z = \lim_{\delta y \to 0} \frac{x(y + \delta y, z) - x(y, z)}{\delta y}$$

where we are taking x as a function of y and z. (δ is the Greek letter *delta* and signifies 'small change in'.) Although z is kept constant in the differentiation, the value of the differential will usually depend on the value of z at which we are calculating it. The above expression for a partial differential is like that for a normal differential when x depends on y only:

$$\frac{\mathrm{d}x}{\mathrm{d}y} = \lim_{\delta y \to 0} \frac{x(y + \delta y) - x(y)}{\delta y}$$

except that z is present as well.

and the equivalent forms of equations *2.16* and *2.17* are

$$C_V = \left(\frac{\partial U}{\partial T}\right)_V \qquad (2.16a)$$

$$C_p = \left(\frac{\partial U}{\partial T}\right)_p + p\left(\frac{\partial V}{\partial T}\right)_p \qquad (2.17a)$$

2.4 CALORIMETRY*

Calorimetry is the name given to experimental techniques associated with measurement of heat. In most calorimetric experiments, the amount of heat is related to the change of temperature by a heat capacity:

$$Q = C\,\Delta T$$

Any term in the equation may be the unknown, but usually the heat capacity or Q is the quantity to be found. Known amounts of heat are most conveniently generated electrically since measurement of electrical energy is simple and accurate.

In all calorimetric experiments it is necessary to prevent or correct for unwanted flow of heat between the system and its surroundings. Very good thermal isolation can be difficult to achieve and how much trouble one has to take depends on the experiment concerned; sometimes simple lagging with a bad conductor like cotton wool or expanded polystyrene is sufficient. Better isolation can be obtained by suspending the system in a vacuum; with no surrounding medium, no heat can be lost by conduction or convection. Heat transfer by radiation (chapter 6) will still take place but this can be reduced by silvering the appropriate surfaces so that they radiate and absorb poorly. In sophisticated calorimetric experiments the system may be surrounded by a container which is kept as nearly as possible at the same temperature as the system so that there is no temperature difference to produce heat flow. This can be done, for example, by detecting any temperature difference with a thermocouple which has one junction on the system and one on the container; the thermocouple e.m.f. can then be used to control (via suitable electronics) the temperature of the container by supplying an appropriate amount of power to a heater attached to it.

* Calorimetry means the measurement of *caloric*, the old name used for heat when it was thought to be a material substance.

If, having taken reasonable precautions, thermal isolation is still not good enough, then corrections have to be made for heat exchanged with the surroundings. For example, if an experiment involves heating a system above the temperature of its surroundings, the heat flow to the surroundings may be found by switching off the supply of heat and observing the rate of fall of temperature. For *sufficiently small* temperature differences between system and surroundings, the heat flow is *always* proportional to the temperature difference, *whatever the means of heat transfer*. (This simple proportionality is known as *Newton's law of cooling*.)

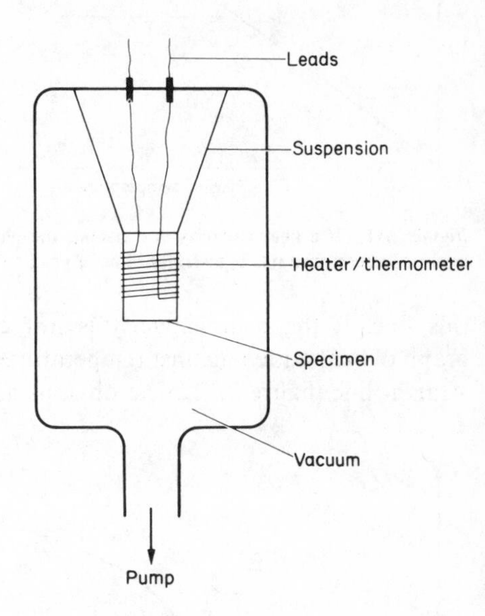

Figure 2.13 The essential elements of a simple vacuum calorimeter

Figure 2.13 illustrates the principles of a simple *vacuum calorimeter*. The sample is suspended on fine threads of a badly conducting material such as nylon. Heat is supplied electrically. The electrical leads to the heater, which could also be used as a resistance thermometer, have to be thin to reduce heat conduction to the surroundings, but not so thin that generation of heat by the current flowing in them is excessive. Clearly a high resistance heater, so that low currents are involved, is an advantage. For accurate measurement of power supplied (or resistance), separate current and potential leads should be used ('four-terminal' connection) so that the measured potential difference should not include voltages developed in the leads as a result of their resistance (figure 2.14).

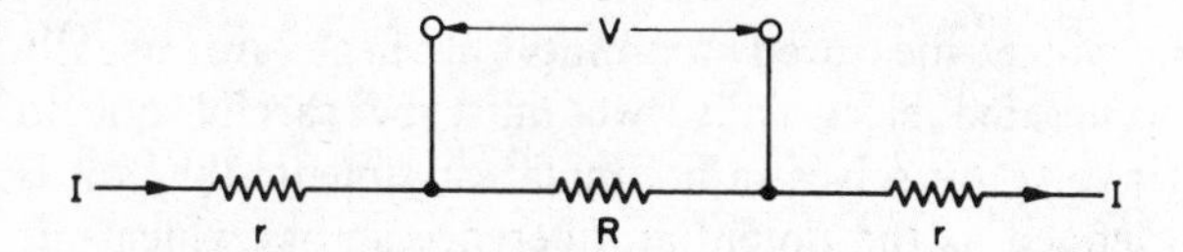

Figure 2.14 Four terminal connection. R is the resistance to be measured and the r's represent resistance in the leads. If separate connections are made directly to R so as to measure the potential difference V across it, then $V = IR$. The potential difference developed across the leads carrying the current includes a contribution from the r's which then need to be known if R is to be found.

In *constant flow calorimeters*, heat is supplied at a constant rate to a fluid which flows steadily through the apparatus, and the temperature rise of the fluid between input and output is measured. Usually the pressure of the fluid does not change significantly between input and output so that, when heat losses are negligible, the energy balance between heat supplied and energy carried away by the fluid gives

$$P = c_p m \, \Delta T \qquad (2.20)$$

where

P is power supplied (watts)
c_p is the specific heat capacity of the fluid
m is the mass of fluid passing per second
ΔT is the temperature rise.

The constant flow technique is useful for finding the heat capacities of gases. Figure 2.15 shows a suitable arrangement. The vacuum flask surrounds the region where the energy is transferred to the flowing gas so that heat loss to the surroundings is small. The volume flowing per second and also the density have to be measured to calculate m. Constant flow calorimetry is also useful for determining the 'calorific value' of fuels (the amount of heat produced by burning a given volume or a given mass). In this case, the fuel is burned at a steady rate and the temperature rise measured in a fluid whose heat capacity is known. The equation is solved for P.

A simple calorimeter for finding the *latent heat of vaporization* of a liquid is illustrated in figure 2.16. (Latent heat of vaporization is the heat required to convert a substance from liquid to vapour. See section 5.4.) A heater boils the liquid and the vapour produced is condensed and collected. When conditions are steady, energy balance gives

$$P = lm$$

where

P is power supplied to heater
l is specific latent heat (heat required to vaporize unit mass)
m is mass vaporized per second.

Microcalorimetry is the name given to the measurement of very small rates of generation of heat such as occur in slow chemical reactions or certain biological systems. In one method, identical containers are suspended in a vacuum space, one containing the material under examination and the other empty. Heat is supplied to the empty container to keep it at the same temperature as the one containing the specimen. (Temperature differences may be detected by a thermocouple.) Since the temperatures of the containers are the same, their heat losses must be

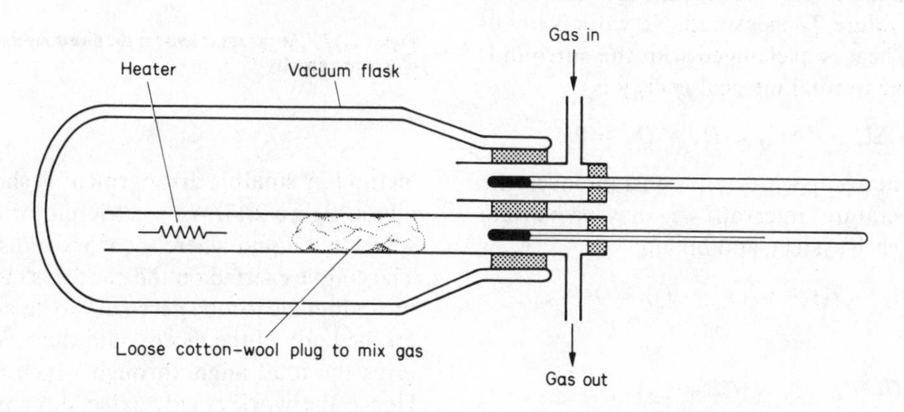

Figure 2.15 A simple constant flow calorimeter for measuring the specific heat capacity at constant pressure of a gas

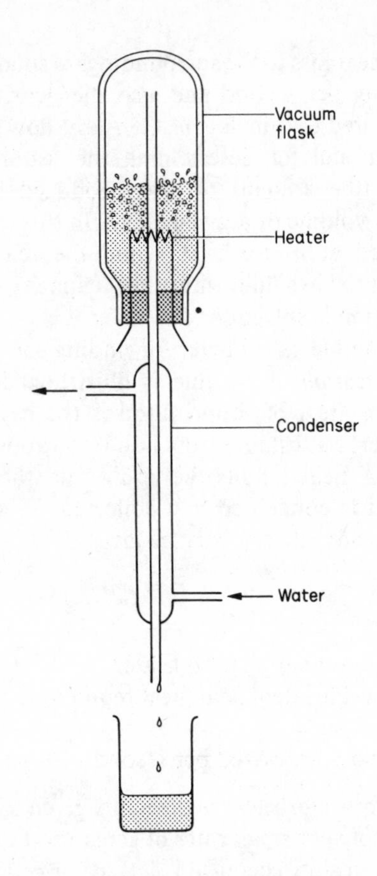

Figure 2.16 A simple calorimeter for determination of latent heat of vaporization

identical, and the heat supplied to the empty container must equal that generated in the other.

The *method of mixtures* is a technique for comparing heat capacities. Two systems, initially at different temperatures T_1 and T_2 are placed in thermal contact (perhaps mixed literally if one of them is a fluid), and the final temperature T_F measured. No work is involved, and if no heat is exchanged with the surroundings, the change in total internal energy is

$$\Delta U = \Delta U_1 + \Delta U_2 = Q_1 + Q_2 = 0$$

Provided the heat capacities are constant over the relevant temperature interval, we may substitute $Q = C\,\Delta T$ for each system, and obtain

$$C_1(T_F - T_1) + C_2(T_F - T_2) = 0$$

or

$$C_1 T_1 + C_2 T_2 = (C_1 + C_2)T_F$$

(Because $Q_1 + Q_2 = 0$ it is sometimes said that 'heat is conserved' in the method of mixtures.)

Since measured amounts of heat are usually generated electrically, we must be careful not to involve ourselves in a circular argument: the volt is defined as the potential difference across which one ampere does work at the rate of one watt, and the simplest means of determining the rate of working is by measurement of electrical heating in a resistor! The joule is defined in mechanical terms, so to avoid circular argument, we should be able to determine at least one heat capacity by a purely mechanical

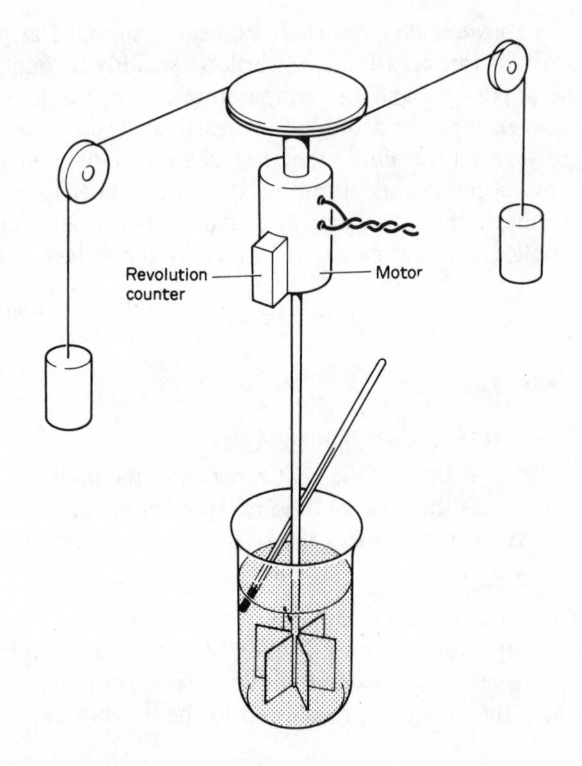

Figure 2.17 An arrangement for the direct mechanical determination of a heat capacity

method. A suitable arrangement is shown schematically in figure 2.17. An electric motor drives a paddle wheel in a liquid where the work is dissipated as heat. The couple exerted on the paddle is measured directly by finding the torque necessary to balance the reaction on the body of the motor, and the revolution counter gives the total angle through which the torque acts. Hence the work is calculated directly. No electrical measurement is involved. The electric motor is simply a convenient device for providing a steady torque.

2.5 FLOW PROCESSES

The first law of thermodynamics amounts to a restatement of the principle of conservation of energy in a form appropriate for thermal processes. Its application to systems involving the steady flow of a fluid leads to interesting results.

Consider any device through which a fluid is flowing steadily (figure 2.18). The flow is said to be *steady* if the

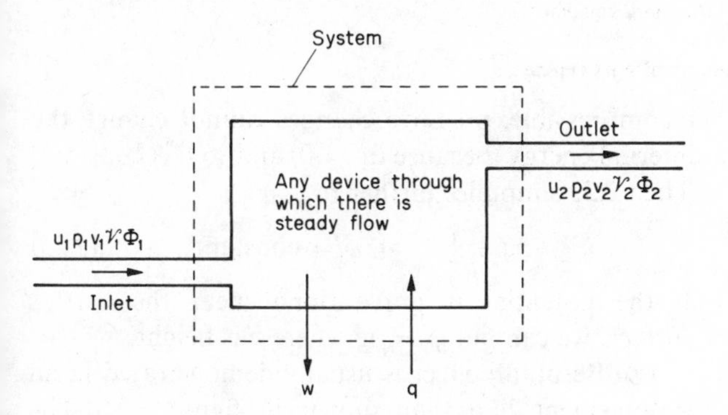

Figure 2.18 A steady flow process

conditions (pressure, velocity, temperature, etc.) at each point in space are constant in time. We apply conservation of energy to the volume within the dotted line. At the inlet, let

u_1 = specific internal energy

p_1 = pressure

v_1 = specific volume

$\mathscr{V}_1$ = speed of flow

ϕ_1 = potential (e.g. gravitational).*

(Remember, specific means per unit mass.)
During the time that unit mass flows through the device, let

q = heat absorbed from surroundings

w = work done on surroundings.

As unit mass enters, the energy carried into the device will be internal (u_1) plus kinetic ($\frac{1}{2}\mathscr{V}^2$) plus potential (ϕ_1):

$$u_1 + \tfrac{1}{2}\mathscr{V}_1^2 + \phi_1$$

However, this is not the total energy entering at the inlet, for some mechanism, perhaps a piston, must be pushing the fluid forward as it goes in so as to keep the input pressure steady at p_1. There is therefore also direct mechanical work (force × distance = pressure × volume) done on the device by the source of fluid, the fluid acting as a mechanical link to communicate this action from source to device. As unit mass flows, the fluid moves forward a volume v_1, and the pressure p_1 is constant, so the work done is $p_1 v_1$. This contribution is known as *flow work*. Thus the total energy entering the device at the inlet as unit mass flows is

$$u_1 + p_1 v_1 + \tfrac{1}{2}\mathscr{V}_1^2 + \phi_1$$

A similar term will apply at the outlet where the flow work corresponds to the device having to do work to force the fluid out against the pressure p_2. Applying conservation of energy (the first law) to the region within the dotted line

$$w = (u_1 + p_1 v_1 + \tfrac{1}{2}\mathscr{V}_1^2 + \phi_1) - (u_2 + p_2 v_2 + \tfrac{1}{2}\mathscr{V}_2^2 + \phi_2) + q \qquad (2.21)$$

This is the general equation for steady flow processes.

In the *constant flow calorimeter* we discussed in the last section (page 27), $w = 0$ and kinetic and potential terms are usually negligible. Also the flow of the fluid is normally relatively unimpeded, so $p_2 \approx p_1$.* Then *2.21* simplifies to

$$q = \Delta u + p\,\Delta v = c_p\,\Delta T$$

where we have used *2.17* for c_p. This is essentially the same result as *2.20* but referred to the flow of unit mass instead of to unit time.

In a *jet engine* we want to maximize thrust by ejecting the exhaust gases with as high a momentum as possible. Potential terms are negligible, and w relatively small since only enough work is extracted from the engine internally to power auxiliary equipment in the aircraft. q represents heat supplied by burning fuel which heats the gases to a very high temperature, raising u and p considerably. At the end of the combustion chamber, $u + pv$ is high but the speed still relatively small. From the nozzle at the back of the engine the hot gases expand. During the expansion, the pressure drops back towards atmospheric, the temperature falls as the gases expand, so $u + pv$ drops and the speed $\mathscr{V}$ rises (figure 2.19). The object is to convert as much of q as possible into kinetic energy so that the change of momentum of the

* ϕ is the Greek letter *phi*.

* The symbol $\approx$ means 'is approximately equal to'.

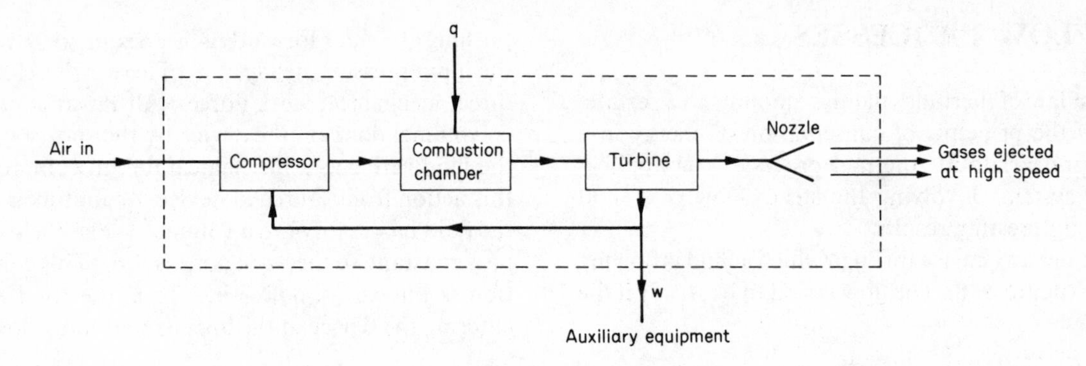

Figure 2.19 The elements of a jet engine

air passing through the engine, and therefore the thrust on the aircraft, should be as large as possible.

We may also apply *2.21* to *streamline flow*. Consider a fluid flowing without turbulence (this is what streamline means) through an area A (figure 2.20).

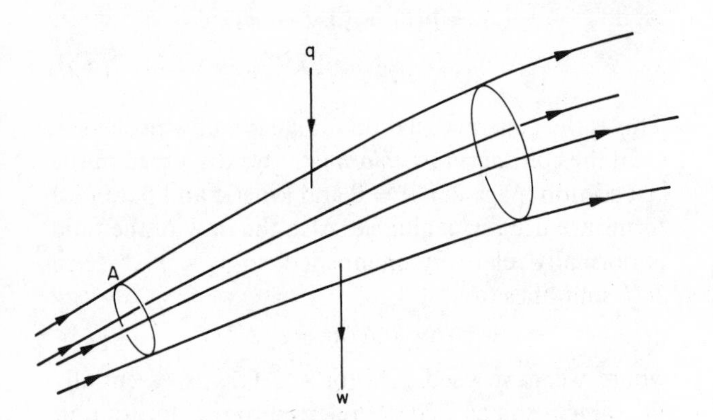

Figure 2.20 Streamline flow as a steady flow process

The gas molecules passing the edges of this area trace out a *tube of flow*, and *2.21* may be applied to any length of this tube (because no fluid crosses the walls of the tube). (The pattern of flow might be set up by making the fluid flow through a rigid tube of appropriate shape, but the existence of such a real tube is not essential to the argument.) The only way in which work can be done across the boundary of the tube is through the fluid's viscosity if the fluid is viscous (page 63). So w is viscous work and q is heat entering. If both these are negligible, *2.21* simplifies to

$$u + pv + \tfrac{1}{2}\mathscr{V}^2 + \phi = \text{constant} \qquad (2.22)$$

The constancy of this quantity along a streamline is known as *Bernoulli's theorem*. The variation of pressure with speed of flow which this equation predicts is known as the *Bernoulli effect*. If the fluid is incompressible, pressure changes cannot change the internal energy (because $dv = 0$) and so u is constant. Then *2.22* simplifies further to

$$pv + \tfrac{1}{2}\mathscr{V}^2 + \phi = \text{constant} \qquad (2.23)$$

If the potential is gravitational near the earth's surface, we can put $\phi = hg$ where h is height.

The Bernoulli effect is usually demonstrated in an arrangement like that shown in figure 2.21. The

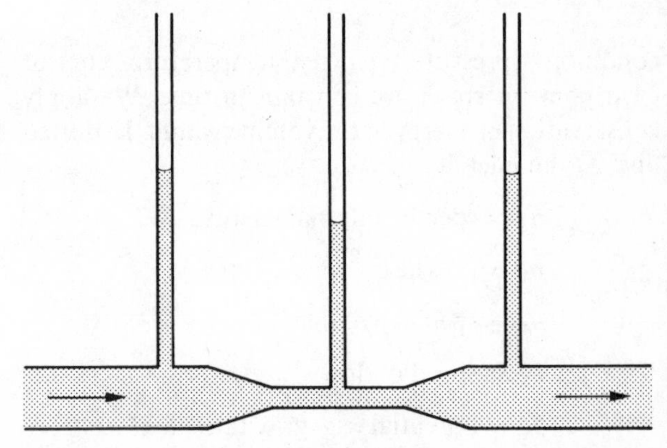

Figure 2.21 Simple demonstration of the Bernoulli effect. Liquid flows through a tube of varying cross-sectional area. Where the tube narrows, the liquid's speed is greater and the pressure drops.

variation of speed along the tube is related to the area by using the *continuity condition*, that is, the fact that mass cannot disappear, so that in the steady state the mass flow must be the same across all cross-sections of the tube. This gives

$$A\mathscr{V}/v = \text{constant} = \text{mass per second}$$

where A is the cross-sectional area of the tube.

Another important application of *2.21* is to *compressors*. Kinetic and potential terms are usually negligible, and if the compression is adiabatic (usually a fairly good approximation), $q = 0$. Then

$$w = \Delta(u + pv)$$

Since $q = 0$, Δu corresponds simply to mechanical work done on the gas in compressing it and is given by equation *2.6*:

$$\Delta u = (p_2v_2 - p_1v_1)/(\gamma - 1)$$

Addition of the flow work contribution $(p_2v_2 - p_1v_1)$ then gives the following for the total work required to drive unit mass through the compressor

$$w = \gamma(p_2v_2 - p_1v_1)/(\gamma - 1) \qquad (2.24)$$

It is a common mistake to forget that compressors not only have to compress, they also have to supply flow work.

2.6 THERMAL CONDUCTIVITY

Heat flows from hot to cold, so if we set up a *temperature gradient* (i.e. temperature varies with position) in a material we expect flow of heat to result. This is a driving force/response situation (like potential difference causing current flow in electricity) in which the temperature gradient acts as the 'force' driving the heat flow. In such cases the response is always linearly proportional to the driving force, provided the latter is not too large. For heat flow, the constant of proportionality relating the rate of heat flow per unit area Φ to the magnitude of the temperature gradient dT/dx is the *thermal conductivity* λ.* Heat flow and temperature gradient are vector quantities, so we write the equation relating them to λ for one component only:

$$\Phi_x = -\lambda \frac{dT}{dx} \qquad (2.25)$$

where the terms have units

Φ_x: $J\,s^{-1}\,m^{-2} = W\,m^{-2}$

λ: $W\,m^{-1}\,K^{-1}$

$\frac{dT}{dx}$: $K\,m^{-1}$.

Φ_x is the rate at which heat flows across unit area normal to the x direction (figure 2.22). The negative sign is present because the heat flows in the direction

* Φ is the capital form of the Greek letter *phi*, and λ is the Greek letter *lambda*.

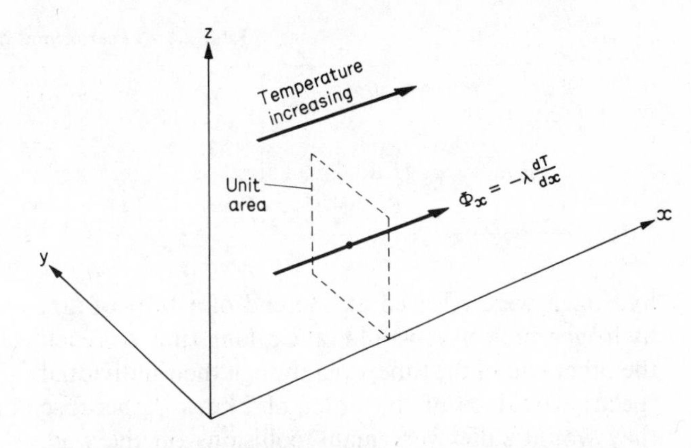

Figure 2.22 The relationship between rate of heat flow, thermal conductivity and temperature gradient

of decreasing temperature, that is, in the direction in which dT/dx is negative. The definition of thermal conductivity is exactly analogous to that of electrical conductivity* σ which is the constant relating the density of current flow J (units: $A\,m^{-2}$) to potential gradient $d\phi/dx$ (units: $V\,m^{-1}$)

$$J_x = -\sigma\frac{d\phi}{dx} = \sigma E_x$$

where we have substituted for the potential gradient in the x direction with the x component of the electric field strength.

In *insulating solids* the mechanism of heat flow is as follows. When a region of the material is warmed, the thermal motions of the atoms increase. The atoms are held in position by forces 'connecting' them to their neighbours, so that the motion of one produces deflecting forces on those next to it. These, in turn, jog their neighbours and so the disturbance travels through the material. But this is exactly how a *sound wave* travels, so we must visualize the thermal motions as spreading out from the heated region in the form of little sound waves, like ripples from a pebble dropped in a pond. (One cannot hear these sound waves. A typical frequency is 10^{13} Hz!) However, we know that heat does not travel over macroscopic distances with the speed of sound (we would often get burnt) and this is because these waves (which are called *phonons*) collide very frequently with each other and with the atoms of the solid. Their mean free path (page 60), the average distance they go before colliding, is only a few nanometres. The thermal energy therefore *diffuses* through the material. An analogy would be what would happen if a little

* σ is the Greek letter *sigma*.

Table 2.1 Experimental values of the Lorenz number

Metal		Ag	Al	Au	Cu	Fe	Pb	Sn	W	Zn
$L/10^{-8}\,\mathrm{V^2K^{-2}}$	at 0 °C	2.31	—	2.35	2.23	2.47	2.47	2.52	3.04	2.31
	at 100 °C	2.37	2.23	2.40	2.33	—	2.56	2.49	3.20	2.33

hydrogen were released at one end of a tube of air: hydrogen molecules would take a long time to reach the other end of the tube, even though their individual speeds would be of the order of $2\,\mathrm{km\,s^{-1}}$, because they would suffer very many collisions on the way. Insulating solids have thermal conductivities of the order of $1\,\mathrm{W\,m^{-1}\,K^{-1}}$.

Compared with insulators, *metals* are relatively good conductors of heat ($\lambda \sim 10^2\,\mathrm{W\,m^{-1}\,K^{-1}}$). The difference results from the fact that, in metals, most of the heat is carried by the electrons. The electrons suffer collisions and are *scattered* just like the phonons are, so the heat still travels by diffusion, but the difference is that the speeds of the electrons are very much greater than the speed of the phonons (i.e. of sound) so that the diffusion process takes place much more rapidly: a typical electron speed is $1000\,\mathrm{km\,s^{-1}}$ while a typical speed of sound in a solid is $4\,\mathrm{km\,s^{-1}}$.

One might expect metals which are good electrical conductors also to be good thermal conductors since the electrons are the carriers in both cases. For the electrical conductivity σ, since the charge on the electron is fixed, we might expect

$$\sigma \propto \text{(constant electrical factors)} \times \text{(ease of motion)}.$$

The thermal conductivity will depend on how much thermal energy the electrons can store and carry with them, i.e. on their heat capacity. So here we might expect

$$\lambda \propto \text{(electron heat capacity)} \times \text{(ease of motion)}.$$

The heat capacity of electrons is proportional to temperature (the reason for this is complicated—it has to do with the Pauli exclusion principle—and you must take it on trust), so this would give

$$\lambda \propto T \times \text{(constant thermal factors)} \times \text{(ease of motion)}.$$

Then we should have

$$\lambda/\sigma T = \text{constant for all metals}$$

This ratio is, indeed, remarkably constant for metals. Its constancy is known as the *Wiedemann–Franz law* and the value of the ratio given by free electron theory is $2.45 \times 10^{-8}\,\mathrm{V^2\,K^{-2}}$, which is called the *Lorenz number*. Table 2.1 shows measured values of the Lorenz number L for several metals.

The usual way of determining thermal conductivity is to supply heat at a known rate, usually electrically, and measure the temperature gradient which results from the flow of that heat through the material. (This is like measuring the resistance of a resistor by forcing a known current through it and measuring the resulting potential difference.) A simple geometry is usually chosen so that the heat flux is easy to calculate. The simplest arrangement is to have the heat flow everywhere parallel. This is the case if the specimen is in the form of a solid cylinder with the heat supplied at one end: if the cylinder is well insulated, the heat flows everywhere parallel to the axis and the total heat flowing is constant along the bar (figure 2.23(a)). With

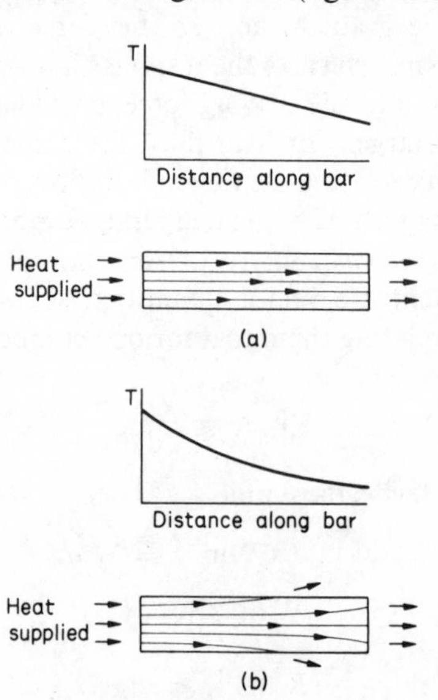

Figure 2.23 Heat flow along a bar. If the sides of the bar are well insulated (a), heat loss through them is negligible, heat flows everywhere parallel to the axis of the bar, and Φ_x is constant along the bar. The temperature then drops linearly with distance. If there is heat loss through the sides of the bar (b), the heat flowing along the bar decreases with distance so that the temperature decreases less rapidly the further one goes from the heated end.

good conductors, to get a large enough thermal resistance to give a reasonable temperature difference without having to use an enormous heat input, one would choose a cylinder with length much greater than its radius (figure 2.24). With a good conductor simple lagging is usually enough to stop significant heat losses to the surroundings.

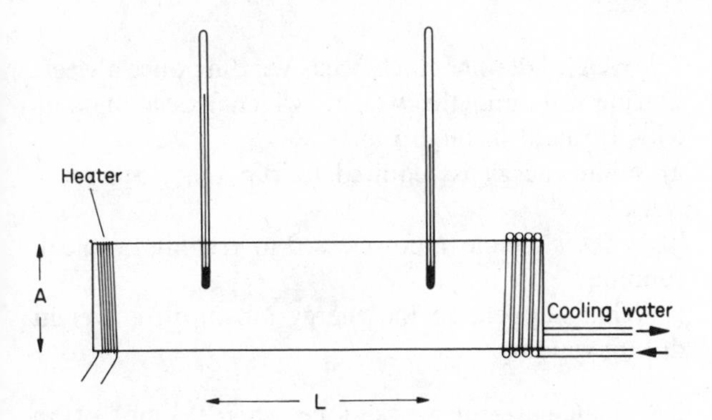

Figure 2.24 A method for measuring the thermal conductivity of a good conductor

A typical arrangement for a bad conductor is shown in figure 2.25. Here, heat flows outwards from a central heater through thin disc-shaped samples of the material to cooled outer plates. Temperatures are measured by inserting thermocouples or other types of thermometer into the heater and outer plates. The

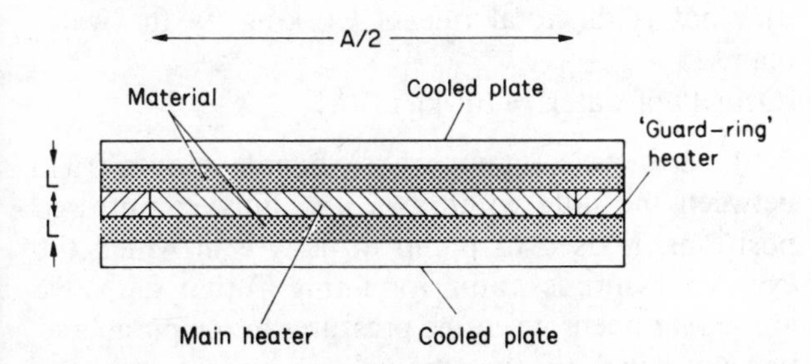

Figure 2.25 A method for measuring the thermal conductivity of a bad conductor

second heater surrounding the central one acts as a 'guard ring', ensuring that all the heat from the central heater flows parallel to the axis of the discs. The two heaters must, of course, be kept at the same temperature. Any heat lost from the sides involves heat from the outer heater and will have little effect on the heat flow from the central one.

With both these arrangements the thermal conductivity λ is related to the total power P flowing through the material and the temperature difference ΔT by

$$P = \lambda A \frac{\Delta T}{L}$$

where A is the total cross-sectional area and L is the length of material across which the temperature difference is measured. (In figure 2.24, L is the distance between the thermometers; in figure 2.25 it is the thickness of the discs.)

An alternative arrangement which is sometimes useful for bad conductors is to use a long cylinder of the material with a heating wire along its axis. The heating wire can act as its own thermometer. The heat flow is radial so that Φ varies with distance from the axis, and the temperature difference between the wire and the outer surface of the cylinder has to be found by integration of *2.25*. Consider the heat flowing out through a length L of a cylindrical shell of radius r and thickness dr (figure 2.26). The surface area of the

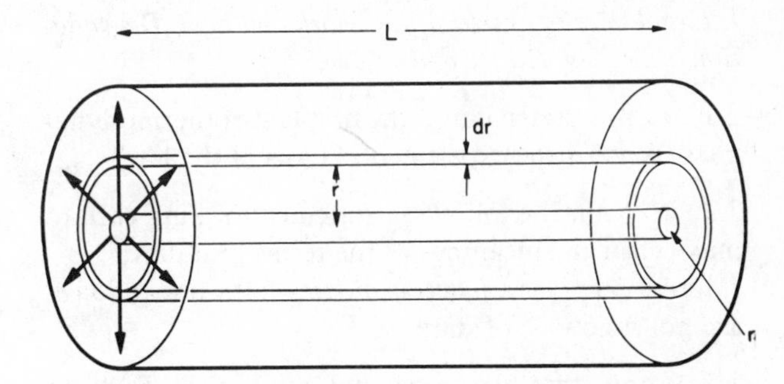

Figure 2.26 Calculation of the case of radial heat flow in a cylinder

cylinder is $2\pi rL$, so the total rate of flow of heat outwards through the cylinder is

$$P = -2\pi rL\lambda \frac{\mathrm{d}T}{\mathrm{d}r} \qquad (2.26)$$

In the steady state there must be continuity of heat flow. (No heat can accumulate anywhere for this would result in changing temperatures.) Therefore P does not vary with radius and must equal the power generated in length L of the heater wire. *2.26* is a differential equation. Rearranging it,

$$P\frac{\mathrm{d}r}{r} = -2\pi L\lambda \, \mathrm{d}T$$

and integrating from the radius of the wire r_1 to the

outer radius of the tube r_2,

$$P \int_{r_1}^{r_2} \frac{dr}{r} = -2\pi L\lambda \int_{T_1}^{T_2} dT$$

giving

$$(P/L) \ln (r_2/r_1) = 2\pi\lambda(T_1 - T_2)$$

where

(P/L) is the power dissipated per unit length of the heater wire

(r_2/r_1) is the ratio of the outer radius of the cylinder to the radius of the wire

and

T_1 and T_2 are the temperatures of the heater wire and of the outer surface of the cylinder.

PROBLEMS

Internal energy, the first law, work and heat. Degradation of energy. Hotter and colder.

2.1 Give a statement of the first law of thermodynamics. What is the experimental basis of the law?

2.2 Explain the following statement, taking care to make clear the meanings of the terms in italics:
Internal energy is a *function of state* but *work* and *heat* are not functions of state.

2.3 Show that the perpetual motion of Escher's waterfall (figure 2.4) violates the first law of thermodynamics. Develop your arguments in two ways:
a) by taking the whole waterfall as a thermodynamic system and applying the first law;
b) by considering in detail the energy changes of a volume of water as it goes around the waterfall.

2.4 Can one body be hotter than another if it is at the same temperature? (Look back at the definitions of temperature and hotness to answer this.)

Some forms of work: by a force, by hydrostatic pressure; electrical work on resistors, capacitors and inductors; work against surface tension; pressure difference across curved surfaces.

2.5 Estimate the maximum rate at which a person does work (a) when running upstairs, (b) during sustained effort. Give your answers in watts and horsepower. [1 hp = 746 W.]

2.6 The elastic of a catapult has an unstretched length of 30 cm and a spring constant of 20 N m^{-1}.
a) What is the work required to stretch the elastic to three times its original length?
b) If 50% of this energy is transferred to a stone of mass 20 g, with what speed is the stone projected?
c) What happens to the rest of the energy stored in the elastic?

2.7 A grandfather clock needs winding once a week. During winding, the weight, which has a mass of 6 kg, is raised through 1 m.
a) What energy is required to run the clock for a week?
b) What is the mean power used in keeping the clock running?
c) What happens to the energy put into the weight during winding?

2.8 Calculate the work done when 0.1 mol of an ideal gas at 300 K expands isothermally from a volume of 2×10^{-3} m^3 to 10×10^{-3} m^3.

2.9 A garden hose will spray water to a height of 8 m. The water flows at 10^{-4} m^3 s^{-1}.
a) What is the potential energy of unit volume of water at the maximum height?
b) How much work must the water mains do on unit volume to project it to this height?
c) What is the pressure of the water supply?
d) What is the total rate of working by the water mains?
[Density of water = 10^3 kg m^{-3}.]

2.10 A bicycle pump, whose handle moves 0.3 m between the fully withdrawn and fully compressed positions, is used to pump up a tyre in which the excess pressure is 2 atm. Assuming (i) that the valve in the tyre opens when the pressures in the pump and tyre are equal, (ii) that the volume remaining in the pump and connector when the pump handle is fully home is negligible, (iii) that the volume in the tyre is large compared with the volume in the pump, and (iv) that the compression is isothermal (in fact it is not; see problems **3.10** and **3.15**), draw a graph with labelled axes showing how the pressure in the pump varies with the position of the pump handle as it is pushed in from the fully withdrawn position.

If the area of the piston in the pump is 300 mm^2, calculate the work done during one pumping stroke.

2.11 A system consists of a battery of e.m.f. E in

series with a capacitor of capacitance C which is initially uncharged. Find the work required to pass a charge Q through the system.

2.12 A 12 V car headlamp bulb draws a current of 5 A. What is the rate at which the battery does work on the bulb? Why does the internal energy of the bulb not increase?

2.13 A capacitor of capacitance 1 μF, which is initially uncharged, and a resistor of resistance 1 MΩ are connected in series. The capacitor is charged by connecting the series combination to a battery of negligible internal resistance and e.m.f. 100 V.
a) Which component stores energy, and which component dissipates energy?
Calculate,
b) the work done by the battery,
c) the work done on the capacitor,
d) the work done on the resistor.
The battery is now disconnected and the capacitor discharged by connecting the resistor directly across its terminals. Give an account of the energy changes which occur.

2.14 An oscillatory circuit consists of a capacitor and an inductor in parallel. The charge on the capacitor Q is given by $Q = Q_0 \cos \omega t$. Work out
a) the instantaneous energy stored in the capacitor,
b) the instantaneous energy stored in the inductor,
c) the total electrical energy in the circuit.
Show that the total energy is constant only if the frequency of oscillation is given by $\omega^2 = 1/LC$.
[You may wish to use the result $\cos^2 \theta + \sin^2 \theta = 1$.]

2.15 A loop of cotton floats on the surface of some water. A little detergent is dropped onto the water surface inside the loop, and the loop opens out and becomes circular. Explain why this happens.

Draw a diagram showing the forces acting on a short length of the circumference of the loop after the detergent has been added. Explain how the short length is in mechanical equilibrium.

2.16 Explain why a soap film collapses if a small hole appears in it.

2.17 Explain why, when a bubble is being blown on the end of a tube, the pressure first rises, then reaches a maximum and then decreases as the bubble grows.

A short length of glass tube with a tap in the centre has a soap bubble on each end.
a) If the bubbles are of equal radius, is the equilibrium stable or unstable when the tap is opened?
b) Describe in detail what will happen when the tap is opened if the radii are different.

2.18 Show that the pressure difference between the inside and the outside of a bubble of radius r is $4\gamma/r$ where γ is the surface tension of the liquid from which the bubble is blown.

2.19 A bubble of radius 10 cm is blown from the end of a narrow tube with soap solution of surface tension $2{\cdot}5 \times 10^{-2}$ N m^{-1}. What is the work done in stretching the soap film? As the bubble is blown, what other work has to be done? (See problem **3.5**.)

2.20 Two flat surfaces are separated by a thin film of liquid. Explain why it is difficult to pull the surfaces apart.

Heat capacities

2.21 Criticize the following statement.
The heat capacity of a body is a measure of how much heat the body can hold.

2.22 Why do the symbols for heat capacities have suffixes?

C_p is different from C_V. Which is larger and why?

Why is the difference between C_p and C_V often neglected for solids?

2.23 A shower is supplied with hot water by a unit in which the water is heated electrically as it flows through from the water main to the shower. What is the maximum rate at which hot water can be supplied if the heater is rated at 7 kW, the temperature of the water from the main is 5 °C and the temperature of the shower water is to be 45 °C?
[Specific heat capacity of water = 4.2 kJ K^{-1} kg^{-1}.
Density of water = 1 Mg m^{-3}.]

2.24 At very low temperatures the molar heat capacity of copper is given by

$$C_{pm}/\text{kJ K}^{-1}\text{ mol}^{-1} = 1.94\,(T/\Theta)^3$$

where $\Theta = 348$ K is the 'Debye temperature' of copper.
a) How much energy is required to heat 100 g of copper from 4 K to 20 K?
If this heat is supplied at a steady rate of 10 mW,
b) what is the total time taken to warm the copper?
c) what is the initial rate of rise of temperature?
d) sketch the form of the temperature/time graph.
[Relative atomic mass of copper = 63.6.]

Principles of calorimetry; heat loss and Newton's law of cooling. Vacuum and constant flow calorimeters. Microcalorimetry. Measurement of latent heats. Method of mixtures.

2.25 What is *calorimetry*? Discuss the physical principles involved in calorimetric measurements, and explain the experimental difficulties encountered when trying to make accurate measurements. Illustrate your discussion by describing any one calorimetric experiment in detail.

2.26 A 10 W electric heater of negligible thermal capacity is inserted into an unlagged metal block which is initially at room temperature. When the heater is switched on, the block warms up and eventually reaches a steady temperature. The heater is then switched off and the block cools at an initial rate of 1.5 °C min^{-1}.

a) What is the heat capacity of the metal block?

b) What was the initial rate of rise of temperature when the heater was first switched on?

2.27 A solid copper cylinder, 50 mm long and of 10 mm radius, is suspended in a vacuum calorimeter. Wound on the cylinder is a length of fine copper wire which is used as heater and resistance thermometer. Initially the resistance of the heater is 100.2 Ω. A current of 100 mA is then passed for 5 min, and then, when conditions are steady, the resistance of the heater is found to be 102.5 Ω.

a) What is the specific heat capacity of copper?

b) What assumptions are you making?

c) What is the most important factor limiting the accuracy of the experiment?

d) Why was it necessary to wait for conditions to become steady?

[Temperature coefficient of resistance of copper = $4.1 \times 10^{-3}\ K^{-1}$. Density of copper = 8.93 Mg m^{-3}.]

2.28 A quantity of water in a beaker of negligible thermal capacity is cooled to a few degrees below freezing point. The beaker is then placed in a warm room, and the times recorded at which it is at various temperatures as it gradually warms. The observations were:

temperature/°C:	−3.0	−2.0	−1.0	0 ...
time/min:	0	0.93	1.89	2.92 ...

temperature/°C:	0	1.0	2.0	3.0
time/min:	169.72	171.84	174.04	176.34

a) Explain the general form of the experimental results.

b) What *quantitative* results can you deduce from the observations?

2.29 In an experiment to determine the variation with temperature of the specific heat capacity of gadolinium between 150 K and 340 K, a sample weighing 51.4 g was suspended in a vacuum calorimeter and heated at a steady rate of 150 mW. The times t between successive temperatures T are shown in the table below. Assuming that the heat capacities of the heater and thermometer were negligible, calculate how the specific heat capacity varies with temperature and plot a graph of your results.

T/K:	150		160		170		180		190		200
t/min:		12.1		12.6		13.1		13.6		14.0	

T/K:	200		210		220		230		240		250
t/min:		14.6		15.1		15.7		16.1		16.9	

T/K:	250		260		270		280		290		295
t/min:		17.8		18.9		20.6		22.7		10.4	

T/K:	295		300		310		320		330		340
t/min:		8.1		14.9		13.7		12.7		12.0	

2.30 A constant flow calorimeter is used to determine the calorific value of methane. Water flows through the calorimeter at $10^{-5}\ m^3\ s^{-1}$ and it is warmed by 15.3 K when the methane is being burned at the rate of 0.70 g min^{-1}. What is the calorific value of methane? [Specific heat capacity of water = 4.18 kJ $K^{-1}\ kg^{-1}$.]

2.31 An aluminium electric kettle weighing 0.5 kg and fitted with a 2.6 kW heating element contains $1.7 \times 10^{-3}\ m^3$ of water at 10 °C.

a) How long will it take for the kettle to come to the boil?

b) How long will it take for the kettle to boil dry?

[Specific heat capacity of aluminium = 99 J $K^{-1}\ kg^{-1}$.
Specific heat capacity of water = 4.19.J.$K^{-1}\ g^{-1}$.
Latent heat of vaporization of water = 2.3×10^6 J kg^{-1}.]

2.32 Some liquid nitrogen (boiling point 77.4 K) is contained in a vacuum flask. Submerged in the liquid is a small electrical heater of resistance 100 Ω. With no power supplied to the heater, the nitrogen is found to boil off at the rate of 0.14 mg s^{-1}. With 5 V applied to the heater, the boil-off rate is 1.40 mg s^{-1}.

a) Why does the nitrogen boil with no electrical heating?

b) How would you measure such small boil-off rates?
c) Calculate the specific latent heat of vaporization of nitrogen.

2.33 A vacuum flask of negligible thermal capacity contains liquid nitrogen at 77.4 K (its normal boiling point). The flask is connected to a pump which reduces the pressure so that the nitrogen boils. Estimate the proportion of the nitrogen which must be boiled off before solid nitrogen begins to solidify from the liquid.
[For nitrogen, latent heat of vaporization = $199\ \mathrm{J\,g^{-1}}$,
Specific heat capacity of the liquid = $2.0\ \mathrm{J\,g^{-1}}$,
Freezing point = 63.3 K.]

2.34 $7 \times 10^{-2}\ \mathrm{m^3}$ of hot water at 60 °C are run into a bath. How much cold water at 7 °C has to be added to lower the temperature to 42 °C? (Neglect the heat capacity of the bath.)

2.35 2 kg of water at 80 °C were poured into a 5 kg, well insulated, copper vessel which was initially at 20 °C. If the final temperature was 69 °C, what is the ratio of the specific heat capacities of copper and water?

2.36 A well insulated copper beaker weighing 150 g contains 10 g of water at 15 °C. 50 g of water at 90 °C are poured into the beaker and the final temperature is 65.7 °C. What is the specific heat capacity of copper?
[Specific heat capacity of water = $4.18\ \mathrm{kJ\,K^{-1}\,kg^{-1}}$.]
(*Note*: the advantage of using a copper vessel for calorimetric experiments is that the high thermal conductivity of copper helps the system to reach a uniform temperature quickly.)

2.37 $10\ \mathrm{cm^3}$ of lead at 290 K are dropped into liquid nitrogen at 77.4 K. When equilibrium is established, 14.3 g of liquid have been boiled off. What is the mean specific heat capacity of lead over this temperature interval?
[Density of lead = $11.4\ \mathrm{Mg\,m^{-3}}$;
Specific latent heat of vaporization of nitrogen = $199\ \mathrm{J\,g^{-1}}$.]

2.38 A copper cylinder of mass 2.0 kg and radius 30 mm is attached coaxially to the horizontal shaft of an electric motor. A cord is wound a few times around the cylinder. One end is attached to a spring balance and the other supports a mass of 0.1 kg. The electric motor is run at 3000 rpm for one minute, during which time the spring balance registers 6.9 kg. At the end of the minute the temperature of the copper is found to have risen by 47 K. Estimate the specific heat capacity of copper. What assumptions are you making?

Flow processes: flow work, constant flow calorimeter, jet engines. Streamline flow, Bernoulli effect. Compressors.

2.39 Show how the principle of conservation of energy applied to a flowing fluid leads to Bernoulli's equation. Explain carefully any assumptions you make.

Water flows through the apparatus shown in figure 2.21 at $3 \times 10^{-5}\ \mathrm{m^3\,s^{-1}}$. The cross-sectional areas of the horizontal tube are $300\ \mathrm{mm^2}$, $100\ \mathrm{mm^2}$ and $300\ \mathrm{mm^2}$. Calculate the difference of levels in the manometers.

Why do the outside manometers register the same pressure even though the water has had to pass through a constriction between them?

2.40 A bucket of water is filled to a depth of 30 cm and there is a small hole in the bottom.
a) Draw a diagram showing how the water flows towards and out through the hole.
b) What is the pressure in the liquid a little way above the hole (say several hole diameters from the bottom of the bucket)?
c) What is the pressure in the water jet a little way below the bucket?
d) Apply Bernoulli's equation to calculate the speed with which the water leaves the bucket.

2.41 A constant flow calorimeter is used to measure the specific heat capacity at constant pressure of air. Measurements are taken with three different flow rates, in each case the power supplied to the heater being adjusted to give the same temperature rise of 30 K. The results were

flow rate/$10^{-6}\ \mathrm{m^3\,s^{-1}}$:	5	10	15
power/mW:	238	421	608

a) Why are measurements made at more than one flow rate?
b) How would you measure the flow rate?
c) What is the specific heat capacity at constant pressure of air?
d) Discuss the various sources of error in such an experiment and estimate what accuracy you might hope to achieve with typical laboratory equipment.

[Density of air under the conditions of measurement $= 1.22\ \text{kg m}^{-3}$.]

2.42 Explain what is meant by *flow work*. Show that there is no flow work for an isothermal flow process with an ideal gas. (An ideal gas obeys $pV/T =$ constant.)

An ideal compressor takes in atmospheric air at density $1.23\ \text{kg m}^{-3}$ and delivers compressed air at 10 atm pressure at a rate of $0.01\ \text{m}^3\ \text{s}^{-1}$. If the compression is isothermal,
a) find the rate at which atmospheric air is taken in by the compressor,
b) find the work which is required to compress unit mass,
c) find the power required to drive the compressor.
d) How does your answer to (c) depend on the density of the atmospheric air?
[*Notes*: (a) In real compressors, the compression is close to adiabatic. For a more realistic calculation, see problem **3.16**. (b) In problems like this, where magnitudes of physical quantities are given at the beginning, you should *always work through the calculation in symbols and substitute magnitudes at the end*. This enables one to check for dimensional consistency and the physical reasonableness of each result as the calculation goes on. It is therefore a valuable means of picking up errors in the physics and slips in the algebra.]

Thermal conductivity: Temperature gradients and heat flow. Mechanisms of heat conduction in insulators and metals, Wiedemann–Franz law. Principles of determination of thermal conductivities for good and bad conductors.

2.43 Describe the mechanisms by which heat flows in solids and explain why metals are better thermal conductors than insulators.

2.44 At very low temperatures the reflection of phonons at a boundary between different materials makes the boundary behave like a thermal resistance so that for heat to flow across it a temperature difference must be present.

When a thin film of gold is deposited on a sapphire crystal, the power conducted at very low temperatures across unit area from gold to sapphire P is given approximately by

$$P/\text{mW mm}^{-2} = 0.14\,(T_{\text{Au}}^4 - T_{\text{S}}^4)/\text{K}^4$$

where T_{Au} and T_{S} are the temperatures of the gold film and sapphire.
a) What is the maximum power that may be generated in a gold film of area $2\ \text{mm}^2$ on sapphire at 4 K if the temperature of the gold film must not rise above 5 K?
b) If the film is heated electrically and it has a resistance of 50 Ω, what is the maximum current that may be passed through it?

2.45 Heat is supplied at a steady rate of 10 W to one end of a well-lagged aluminium cylinder the other end of which is water-cooled. The cylinder has a radius of 20 mm and two thermometers 100 mm apart show a temperature difference of 3.8 °C. What is the thermal conductivity of aluminium?

2.46 Estimate the rate at which heat would be lost on a cold winter's day through $30\ \text{m}^2$ of solid brick wall 30 cm thick. [For brick, $\lambda \approx 0.5\ \text{W m}^{-1}\ \text{K}^{-1}$.]

2.47 Expanded polystyrene is often used as a thermal insulator for buildings. With what thickness of expanded polystyrene would a 300 mm solid brick wall have to be covered to halve the heat loss?
[Thermal conductivities of expanded polystyrene and brick are about $0.04\ \text{W m}^{-1}\ \text{K}^{-1}$ and $0.5\ \text{W m}^{-1}\ \text{K}^{-1}$ respectively.]

2.48 How many 2 kW electric heaters would be needed to supply the heat which would be lost through the $4\ \text{m}^2$ of window in a typical living room if the glass is 3 mm thick and the temperatures of its inner and outer surfaces are 25 °C and 0 °C respectively? Comment on your result.
[For window glass, $\lambda \approx 1.1\ \text{W m}^{-1}\ \text{K}^{-1}$.]

2.49 At 4 K the thermal conductivity of high purity copper is about $1.5\ \text{kW m}^{-1}\ \text{K}^{-1}$. Calculate the temperature drop when $5\ \mu\text{W}$ are conducted along a copper wire 10 cm long and 1 mm in diameter.

2.50 Ice on a pond is 10 mm thick. A cold wind keeps the upper surface of the ice at −5 °C. How fast is the thickness of the ice increasing? Explain carefully any assumptions you make.
[For ice, $\lambda \approx 2.1\ \text{W m}^{-1}\ \text{K}^{-1}$.
Specific latent heat of fusion of water $= 333\ \text{kJ kg}^{-1}$.]

2.51 Equal lengths of copper and iron wires of the same radius are joined end to end. The free end of the

copper is maintained at 0 °C and the free end of the iron at 100 °C. The whole length is well-lagged so that heat losses are negligible.
a) What is the temperature at the junction between the metals?
b) If a millivoltmeter is connected by copper leads to the free ends of the composite wire, what will its reading be?
[Thermal conductivity of copper = 390 W m^{-1} K^{-1},
thermal conductivity of iron = 63 W m^{-1} K^{-1},
sensitivity of a copper–iron thermocouple =
8.6 μV K^{-1}.]

2.52 A metal pipe of outer radius 10 mm carries steam at 100 °C. It is insulated by a close-fitting sleeve of expanded polystyrene 15 mm thick.
a) What is the rate of loss of heat from a 10 m length of pipe if the surrounding air is at 20 °C?
b) At what rate would steam condense in this length of pipe as a result of the heat loss?
[Thermal conductivity of expanded polystyrene is about 0.04 W m^{-1} K^{-1}.
Specific latent heat of vaporization of water =
2.26 MJ kg^{-1}.]

3 The Ideal Gas

3.1 EXPERIMENTAL LAWS

The ideal gas is one which obeys Boyle's and Joule's laws at all temperatures and pressures. Real gases do not obey these laws at all temperatures and pressures. In particular, an ideal gas does not condense to a liquid because there are no attractive forces between its molecules to make them come together, whereas all real gases condense at sufficiently low temperatures. On the other hand, if the temperature is not too low and the pressure not too high, real gases do obey Boyle's and Joule's laws quite well. It is fortunate that the physicists of the last century did many of their early experiments with gases under conditions such that their behaviour was close to ideal. If they had not, the development of thermal physics would have taken much longer. In the next chapter we shall examine the reasons why real substances behave the way they do. In this chapter we shall mainly be concerned with the properties of ideal gases and with ideas which follow from them.

Boyle's law states:

If the temperature of a given mass of gas is kept constant, then the product pV is constant.

Note that this law does not involve any particular temperature scale since only *constancy* of temperature is referred to. (One may test for constancy with any thermometer however its scale is calibrated.) Figure 3.1 shows a simple experimental arrangement for verifying Boyle's law.

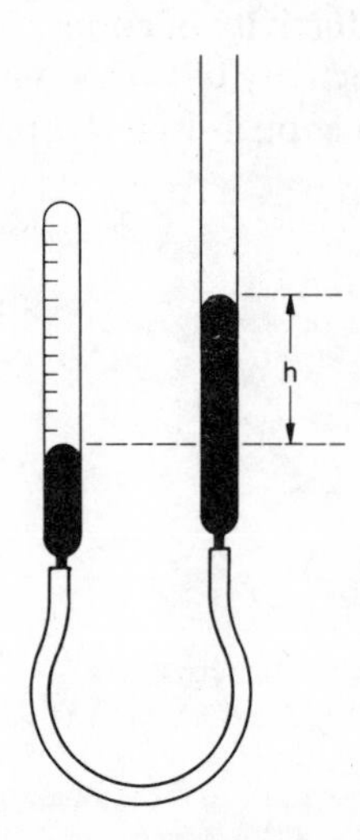

Figure 3.1 A simple arrangement for verifying Boyle's law

Boyle's law means that the product pV can only depend on temperature: for different temperatures the product takes different constant values. The value of the product can therefore be used to define an empirical scale of temperature:

$$pV = a\Theta \qquad (3.1)$$

where a is a constant which may be chosen to adjust the size of the unit of temperature. This is the ideal gas scale which became adopted as the fundamental measure of temperature before the idea of thermodynamic temperature was developed. In chapter 7 we shall show that the scale defined by *3.1* is identical to thermodynamic temperature. Hence, in fact, the equation of state of the ideal gas is

for 1 mol $$pV_m = RT \qquad (3.2)$$

for n mol $$pV = nRT \qquad (3.3)$$

where the *gas constant* R is chosen to set the size of the unit of thermodynamic temperature. Its value is

$$R = 8.31\ \mathrm{J\,K^{-1}\,mol^{-1}}$$

Joule's law states:

*The internal energy of an ideal gas depends only on its temperature.**

Put the other way around, this means that provided the temperature is kept constant, the internal energy does not depend on pressure or volume. This law was based on experiments in which gases were allowed to

* A more formal way of expressing this is to say that the internal energy is a *function* of temperature only.

make a *free expansion* under thermally isolated conditions. A free expansion (also known as a Joule expansion) is one in which no work is done by the gas in expanding. A way of doing this is illustrated schematically in figure 3.2: a container is divided into

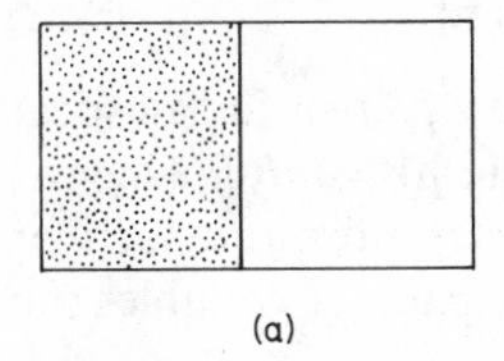
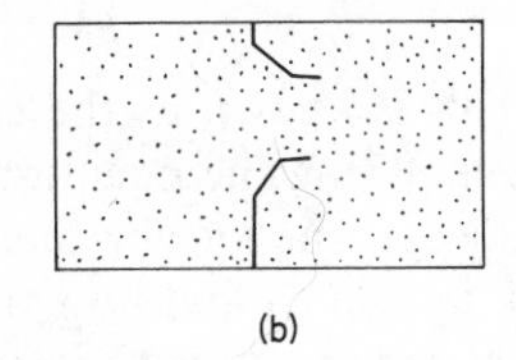

(a) (b)

Figure 3.2 A free expansion. When the partition is ruptured, the gas expands to fill the whole container without doing external work.

two regions by a diaphragm, one containing the gas and the other evacuated; the diaphragm is broken and the gas rushes (violently) through to fill the whole container. Since no piston is pushed back, no work is done. If the system is thermally isolated, no heat is exchanged with the surroundings. Therefore the internal energy remains constant:

$$\Delta U = Q + W = 0$$

What Joule observed was that, when the gas had come to equilibrium after the expansion, its temperature was unchanged. Now suppose that the internal energy *did* depend on volume, as it would, for example, if there were attractive forces between the molecules (see sections 4.7 and 5.1). Then there would be two contributions to the internal energy: one depending on volume (since work must be done to change the distance between molecules when there are forces between them) and one depending on temperature (which is a measure of the energy of thermal motion of the molecules (page 47)). In the free expansion the total energy is constant, so that, if energy is needed for moving the molecules further apart, it can only come from a sharing of the energy in the thermal motions. The temperature of the gas would therefore fall. Since Joule saw no temperature change, he concluded that there could be no part of the total internal energy which depended on volume (or pressure) alone.

It follows from Joule's law that we may write changes in internal energy of an ideal gas in a very simple form. If we warm the gas an infinitesimally small amount $\mathrm{d}T$ at constant volume, we have

$$\mathrm{d}Q = C_V\,\mathrm{d}T$$

which gives

$$\mathrm{d}U = \mathrm{d}Q + \mathrm{d}W = C_V\,\mathrm{d}T \qquad (3.4)$$

since $\mathrm{d}W = 0$. But U *only* depends on temperature, so that changes in internal energy will *always* be given by *3.4* even if the volume does change. Therefore, for *any* change in an ideal gas we must have

$$\mathrm{d}U = C_V\,\mathrm{d}T \qquad (3.5)$$

3.2 ADIABATIC CHANGES

Boyle's law connects changes of pressure and volume when temperature is kept constant. When we compress a gas, we do work on it (which increases its internal energy) so that, if the temperature is to remain constant, the gas has to lose heat to the surroundings. If we compress the gas adiabatically (that is, if we prevent loss of heat) the temperature will rise, and the product pV will not be constant. We may use the first law to find how pressure, temperature and volume are related in adiabatic changes in an ideal gas. First, we derive a formula for the difference of the principal heat capacities C_p and C_V.

If we warm a gas slightly at constant pressure we have

$$\mathrm{d}Q = C_p\,\mathrm{d}T \quad \text{and} \quad \mathrm{d}W = -p\,\mathrm{d}V$$

Substituting in the expression for the change in internal energy,

$$\mathrm{d}U = \mathrm{d}Q + \mathrm{d}W = C_p\,\mathrm{d}T - p\,\mathrm{d}V \qquad (3.6)$$

Differentiating the equation of state *3.3*, keeping pressure constant we have*

$$p\,\mathrm{d}V = nR\,\mathrm{d}T \qquad (3.7)$$

and substituting this in *3.6*

$$\mathrm{d}U = (C_p - nR)\,\mathrm{d}T$$

But we have already shown that for an ideal gas $\mathrm{d}U$ is always given by

$$\mathrm{d}U = C_V\,\mathrm{d}T$$

* If we differentiate with respect to T keeping p constant we get

$$p\frac{\mathrm{d}V}{\mathrm{d}T} = nR$$

from which we conclude that infinitesimal changes in volume and temperature, $\mathrm{d}V$ and $\mathrm{d}T$, are connected by *3.7* above (if p is constant).

Equating these two expressions for dU we obtain

for n mol $\qquad C_p - C_V = nR \qquad (3.8)$

for 1 mol $\qquad C_{mp} - C_{mV} = R \qquad (3.9)$

The significance of the difference in the principal heat capacities was discussed in the last chapter (page 24): if we heat a gas at constant volume the heat goes to increase the thermal energy of the molecules only; if we heat at constant pressure, then for the same change in temperature, *extra* energy is needed to provide work (for pushing the piston back). For an ideal gas the difference comes out to be $nR\,\Delta T$.

We arrived at this result by putting all the changes in the equations in terms of dT. We now derive the relation between p and V in adiabatic changes by deducing how dp and dV are related. This will be a differential equation which we shall then integrate to find the formula connecting p and V themselves.

For an adiabatic change $dQ = 0$, so

$$dU = dW = -p\,dV$$

But from *3.5* we also have

$$dU = C_V\,dT$$

so

$$-p\,dV = C_V\,dT \qquad (3.10)$$

Differentiating the equation of state *3.3* (by parts) we get another formula for dT:

$$nR\,dT = p\,dV + V\,dp$$

Eliminating dT between this and *3.10*

$$-nRp\,dV = C_V(p\,dV + V\,dp)$$

Rearranging this and using *3.8* for nR,

$$pC_p\,dV = -VC_V\,dp$$

$$\frac{C_p}{C_V}\frac{dV}{V} = -\frac{dp}{p} \qquad (3.11)$$

Now the principal heat capacities of an ideal gas are constant (see section 3.5.4) so their ratio is also constant and this equation can be integrated. The symbol γ is always used for the ratio of the principal heat capacities:

$$\gamma = \frac{C_p}{C_V} = \frac{C_{mp}}{C_{mV}} = \frac{c_p}{c_V}$$

Substituting in *3.11*,

$$\gamma\frac{dV}{V} = -\frac{dp}{p} \qquad (3.12)$$

Integrating,

$$\gamma \ln V = -\ln p + \text{constant}$$

$$\ln V^\gamma + \ln p = \text{constant}$$

$$\ln (pV^\gamma) = \text{constant}$$

$$pV^\gamma = \text{constant} \qquad (3.13)$$

This is known as the *adiabatic equation* for an ideal gas. Using the equation of state $pV = nRT$ which is *always* true for *any* change, we can substitute for p or V in *3.13* to find how the other pairs of variables are related in adiabatic changes:

$$TV^{(\gamma-1)} = \text{constant} \qquad (3.14)$$

$$T^\gamma p^{(1-\gamma)} = \text{constant} \qquad (3.15)$$

We should notice that when a gas is compressed adiabatically it gets hotter (equation *3.14*, remembering that $\gamma > 1$) so the pressure rises *more* than it would if the compression were carried out isothermally. This is also obvious from equation *3.13* which gives $p \propto (1/V)^\gamma$ while Boyle's law gives $p \propto (1/V)$; since $\gamma > 1$ this means that pressure depends more strongly on volume in adiabatic changes than in isothermal. We will look at this in a little more detail in the next section.

3.3 BULK MODULUS

Bulk modulus measures how hard it is to compress a material.* It is defined by

$$\text{bulk modulus} = \frac{\text{extra pressure applied}}{\text{fractional change in volume}}$$

In symbols, the bulk modulus K is given by

$$K = -V\left(\frac{\partial p}{\partial V}\right)$$

The negative sign is present because an increase in pressure (dp positive) produces a decrease in volume (dV negative). The modulus is a positive quantity, and, as we see from this equation, has the dimensions of pressure, so that its normal unit is the pascal.

* Bulk modulus is the reciprocal of compressibility, κ, which measures how easily a material is compressed:

$$\text{compressibility} = \frac{\text{fractional change in volume}}{\text{extra pressure required to produce it}}$$

in symbols, $\qquad \kappa = -\frac{1}{V}\left(\frac{\partial V}{\partial p}\right)$

κ is the Greek letter *kappa*.

As usual, however, the relationship between p and V is only fixed if the system is constrained so that the way the change takes place is defined. There are two common cases. The change will be *isothermal* if the gas is in thermal contact with surroundings at constant temperature *and if the compression is slow enough for there to be enough time for the heat to flow between system and surroundings without having to develop any significant temperature difference between them.* We then have the isothermal bulk modulus:

$$K_T = -V\left(\frac{\partial p}{\partial V}\right)_T \qquad (3.16)$$

The differential coefficient $(\partial p/\partial V)_T$ is obtained immediately for an ideal gas by differentiating the equation of state, $pV = nRT$, keeping temperature constant:

$$p\,\mathrm{d}V + V\,\mathrm{d}p = 0$$

which rearranges to

$$\left(\frac{\partial p}{\partial V}\right)_T = -\frac{p}{V}$$

and substituting in *3.16*

$$K_T = p \qquad (3.17)$$

At the other extreme, the compression will be adiabatic if the gas is thermally isolated or if the compression is so fast that there is insufficient time for any significant amount of heat to flow between system and surroundings. We then have the adiabatic bulk modulus:

$$K_S = -V\left(\frac{\partial p}{\partial V}\right)_S \qquad (3.18)$$

where we have used the suffix S to indicate adiabatic. (This convention comes out of thermodynamic theory. S stands for entropy (section 7.9) which is the quantity which remains constant in adiabatic changes.)

This time, the differential coefficient for an ideal gas is given directly by equation *3.12* which relates infinitesimal changes of pressure and volume in adiabatic changes. Rearranging *3.12*

$$\left(\frac{\partial p}{\partial V}\right)_S = -\gamma\frac{p}{V} \qquad (3.19)$$

Alternatively we could derive this expression by differentiating the adiabatic equation, which is a result which should be memorized: Taking logs of equation *3.13*

$$\ln p + \gamma \ln V = \text{constant}$$

and differentiating

$$\frac{\mathrm{d}p}{p} + \gamma\frac{\mathrm{d}V}{V} = 0$$

which is the same as equation *3.12* and immediately gives *3.19*. Substituting *3.19* in the equation for the adiabatic bulk modulus, we get

$$K_S = \gamma p \qquad (3.20)$$

Comparing *3.17* and *3.20* we see that when a gas is compressed adiabatically the pressure increases γ times as fast as when the compression is isothermal.

Equation *3.20* gives a direct way of measuring γ. The speed of sound v_s is given by

$$v_s = \sqrt{K/\rho}$$

where ρ is density.* Compression and rarefaction follow one another so rapidly in a sound wave in a gas that there is no time for heat to flow from compressions (where the gas gets hot) to neighbouring rarefactions (where the gas is cooler). Therefore, the correct modulus is the adiabatic one and

$$v_s = \sqrt{\gamma p/\rho} = \sqrt{\gamma RT/M} \qquad (3.21)$$

where M is molar mass. Thus γ may be determined directly by measuring the speed of sound. In this way, γ may be found quite accurately (say to 1%) with relatively simple apparatus (figure 3.3). C_p can also be measured quite accurately by the constant flow method (page 27); but C_V for a gas is relatively difficult to determine directly with much precision. Consequently the most convenient way of finding both principal heat capacities is usually by measuring C_p and v_s.

3.4 ELEMENTARY KINETIC THEORY

The kinetic theory of gases depends on the atomic nature of matter. Chemical evidence for atomicity is based on the quantitative laws of chemistry: for example, the fact that chemical reactions between gases involve volumes (measured at the same temperature and pressure) which are in simple whole-number ratios to one another (page 48). The most

* ρ is the Greek letter *rho*.

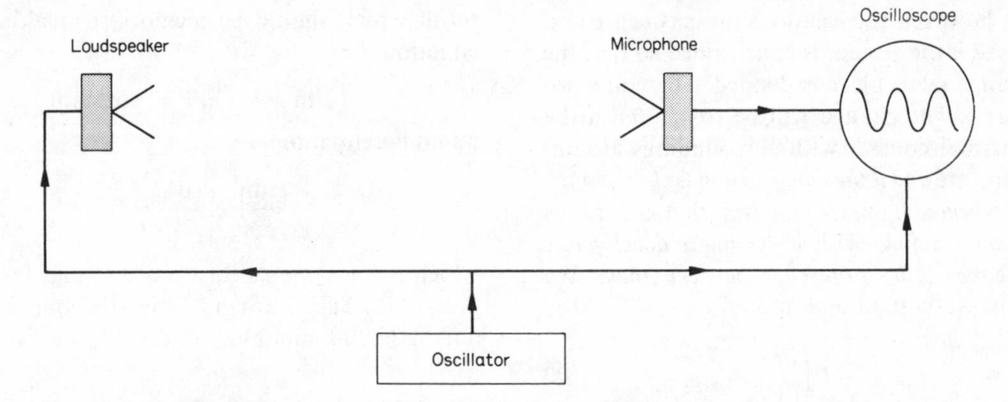

Figure 3.3 A simple arrangement for determining the speed of sound. The oscilloscope sweep is synchronized with the oscillator. As the loudspeaker and microphone are moved further apart, the trace on the screen shifts because of the increasing time delay between transmission and reception. The trace moves by one cycle as their separation is increased by one wavelength, λ. If the oscillator frequency is ν, $v_s = \nu\lambda$.

direct physical evidence for the atomic structure of solids comes from X-ray diffraction experiments in which the solid behaves like a three-dimensional diffraction grating, the pattern of scattered beams depending on the spatial arrangement of the atoms (figures 3.4 and 3.5). The electron microscope, as a result of lens aberrations, cannot quite resolve individual atoms, but the electrons in the beam of the microscope have wave-like properties (as a result of the wave-particle duality of nature) and they are diffracted in the same way as X-rays. In the field ion microscope, where the imaging is done by ions, individual atoms can be 'seen'. The fact that matter retains its particulate structure in the fluid states is demonstrated by *Brownian motion*: very small particles suspended in a fluid are seen to be in continuous irregular movement as a result of being pummelled continually from all sides by molecular impacts. Brownian motion is easily observed under the microscope. Simple laboratory demonstrations can use: in liquids, the smaller fat globules in diluted milk; in gases, the ash particles in cigarette smoke.

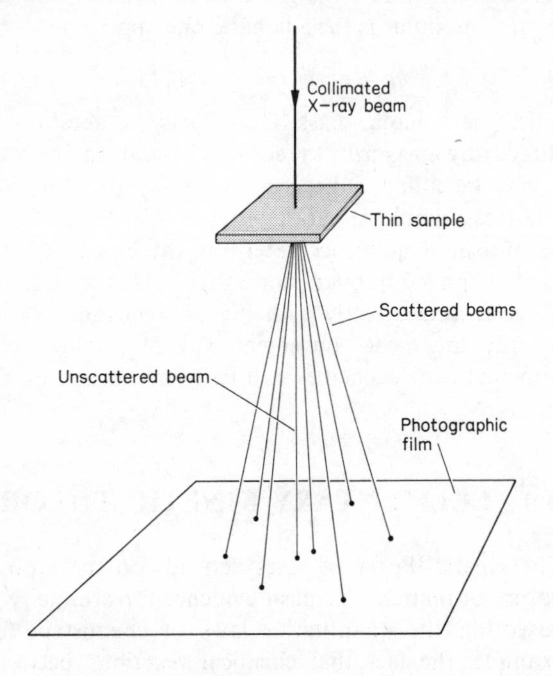

Figure 3.4 X-ray diffraction. The X-rays are reflected by parallel planes of atoms in the sample. If the source produces a broad band of X-ray wavelengths, there will always be some wavelengths present which satisfy the Bragg condition $\lambda = 2d \cos\theta$ for each set of atomic planes, whatever their spacing.

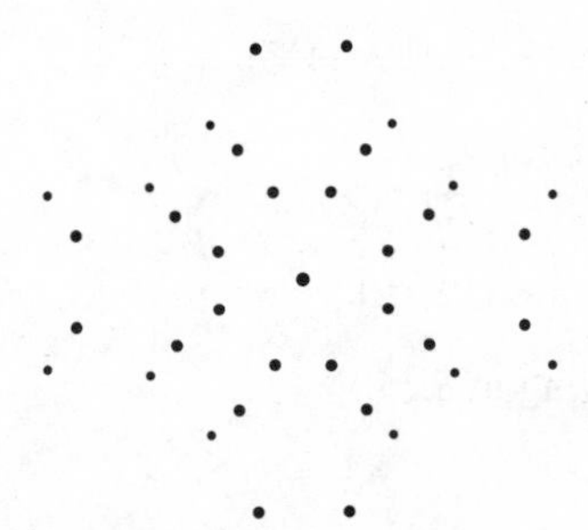

Figure 3.5 The pattern of scattered beams produced by diffraction from a single crystal of diamond

The *basic assumptions* of the kinetic theory of gases are:

1) that gases contain very large numbers of molecules;

2) that the volume occupied by the molecules themselves is small in comparison with the volume of the gas.
3) that the molecules are in continual rapid motion, flying in all directions and colliding with one another and with the walls of the container;
4) that the pressure on the container is the result of the reaction on the walls as the molecules strike and rebound. The impacts are so numerous that the arrival of individual molecules is not normally distinguished: the pressure measures the mean reaction force.

In order to derive a simple expression for pressure, we make several *simplifying assumptions* which amount to making the gas ideal:

1) The volume occupied by the molecules themselves is negligible in comparison with the volume in which they are contained.
2) Intermolecular forces are negligible. If intermolecular forces were not negligible, then the motion of a molecule would be affected by the presence of others close by, and, in particular, as it approached the wall of the container its speed of impact would be reduced because the forces to other molecules would no longer be pulling in all directions but would tend to be directed back into the body of the gas (figure 3.6). As a result, the molecule would hit the wall less hard. This assumption is equivalent to Joule's law: if there were intermolecular forces the internal energy would depend on the volume occupied by the gas (as well as on temperature) because of the change in mutual potential energy as the average distance between the molecules changes.
Assumptions 1 and 2 mean that each molecule moves around the container as if it were the only one present.

3) The collisions with the walls of the container are elastic so that momentum normal to the surface is reversed while that parallel is conserved. This means that the molecules are reflected 'specularly': as a ray of light would be from a mirror.

There are elementary derivations of the expression for pressure in an ideal gas which try to take account of the random directions in which the molecules are moving by supposing that the result is the same as if one third of the molecules were moving parallel to each of the three mutually perpendicular directions. This approach *happens* to give the right answer for pressure because of a mathematical fluke, but it gives the wrong answer to other problems (like the calculation of the number of molecular impacts per area per second). We use an argument which does not involve this false assumption. In the Appendix we give a more rigorous analysis in which the directions of molecular motion are taken into account in detail.

Consider a molecule bouncing around inside a spherical container such that its path between impacts subtends an angle 2θ at the centre (figure 3.7). Let the molecular mass be m and its speed c. The distance between impacts is $2r \sin\theta$ so that in unit time it

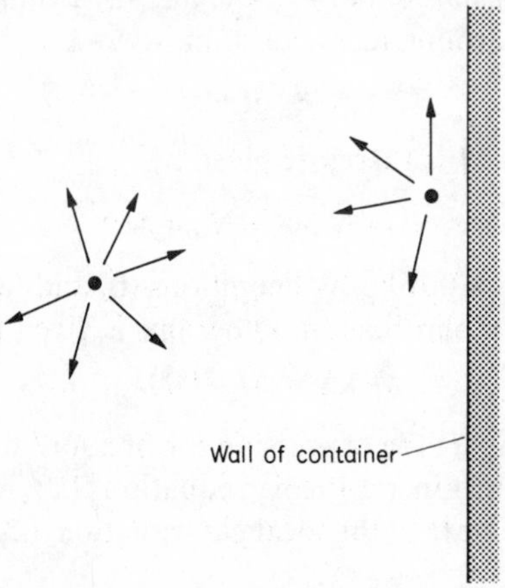

Wall of container

Figure 3.6 Effect of intermolecular forces. The attractive forces to neighbouring molecules, on average, cancel one another out for a molecule deep in the gas because it is surrounded on all sides. But for a molecule close to the wall the average attractive force to other gas molecules is directed back into the bulk of the gas and acts to reduce the molecule's speed before impact.

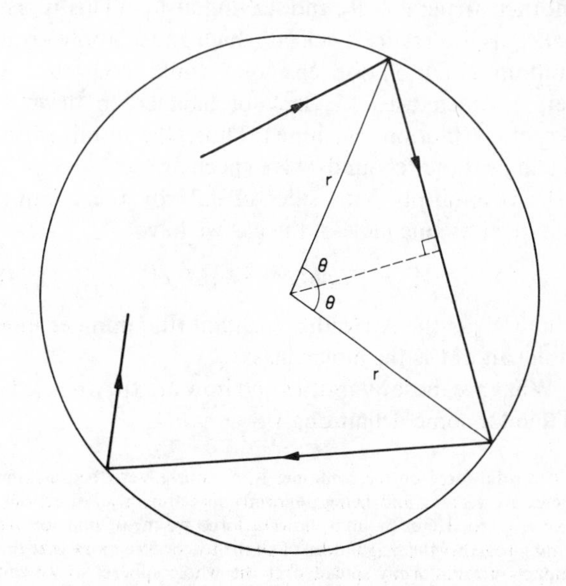

Figure 3.7 Calculation of the pressure exerted by an ideal gas

makes $c/2r \sin\theta$ collisions with the walls. At each collision, its momentum normal to the surface is reversed so that the impulse given to the surface is $2 \times mc \sin\theta$. Then the mean outward force contributed by this molecule is

$$(c/2r \sin\theta) \times (2mc \sin\theta) = mc^2/r$$

which is independent of the angle of impact.

If the density of molecules (number per volume) is n, the total number present is $\frac{4}{3}\pi r^3 n$, and the total outward force* is $\frac{4}{3}\pi r^2 mnc^2$.

But this is distributed over the whole sphere which has an area $4\pi r^2$ so that the pressure (force per area) on the sphere is

$$p = \tfrac{1}{3}mnc^2$$

However, the molecules will not all be moving at the same speed so that the (mean) pressure will in fact be

$$p = \tfrac{1}{3}mn\langle c^2\rangle \qquad (3.22)$$

where $\langle c^2\rangle$ means the average of the square of the molecular speeds. Note that $mn = \rho$, the density, which gives an alternative form

$$p = \tfrac{1}{3}\rho\langle c^2\rangle \qquad (3.23)$$

These results are not affected if we remove the third simplifying assumption, namely that the collisions with the walls are elastic. The molecular motions must be completely random so that for every molecule arriving at θ_1, suffering a diffuse (non-specular) reflection, and leaving at θ_2, there will be another arriving at θ_2 and leaving at θ_1. (This is the *principle of detailed balance* which must apply to a random situation for else one could construct a device which used the lack of balance to drive a perpetual motion machine.) Thus, the total effect is still as if the rebounds were specular.

If we multiply both sides of *3.22* by the volume occupied by one mole of the gas we have

$$pV_m = \tfrac{1}{3}N_A m\langle c^2\rangle = \tfrac{1}{3}M\langle c^2\rangle \qquad (3.24)$$

where N_A is the Avogadro constant (the number in a mole) and M is the molar mass.

What are these quantities and how are they related? (Time for some definitions!)

* The total force on the container is, of course, zero because the forces are vectors, and, being uniformly spread over all directions, sum to zero. Here, by total *outward* force we mean that we are adding together the magnitudes of all the forces. We know that the impacts are uniformly spread over the whole sphere, so we can divide by area to get force per unit area, which is pressure.

a) The *mole* is the amount of substance which contains as many elementary entities (i.e. molecules when we discuss gases) as there are atoms in 0.012 kg of the isotope of carbon, ^{12}C (the most abundant isotope in natural carbon).

b) The *Avogadro constant* N_A is the number of atoms in 0.012 kg of ^{12}C. (i.e. it is the number in a mole.)

The mass of a single molecule m is most directly calculated from the unified atomic mass constant (or the mass of the proton) and the relative molecular mass.

c) *Relative molecular mass*† M_r is the ratio of the mass of the molecule to one twelfth of the mass of an atom of ^{12}C. From (b) it follows that relative molecular mass is numerically equal to the mass in grams of a mole of molecules.

d) *Relative atomic mass*‡ A_r is the ratio of the mass of the atom to one twelfth of the mass of an atom of ^{12}C. From (b) it follows that relative atomic mass is numerically equal to the mass in grams of a mole of atoms.

e) The *unified atomic mass constant* m_u is one twelfth of the mass of an atom of ^{12}C. The unified atomic mass constant is a convenient unit for atomic sized masses.

Thus,

$$m = M_r m_u \qquad (3.25)$$

The *mass of the proton* m_p is equal, to within 1%, to the unified atomic mass constant, so

$$m \approx M_r m_p \qquad (3.26)$$

Molar mass M is therefore given by

$$M = N_A m = N_A m_u M_r$$

But $N_A m_u$ is 0.001 kg by definitions (b) and (e) above. Hence, the molar mass in kilograms is given by

$$M/\text{kg} = M_r/1000 \qquad (3.27)$$

If we compare the expression for pressure which we derived from kinetic theory, equation *3.24*, with the equation of state of the ideal gas, equation *3.2*, we find

$$\tfrac{1}{3}M\langle c^2\rangle = RT$$

Dividing through by N_A,

$$\tfrac{1}{3}m\langle c^2\rangle = RT/N_A = kT \qquad (3.28)$$

† Formerly called 'molecular weight'.
‡ Formerly called 'atomic weight'

where $k = R/N_A$ is called the *Boltzmann constant.* In terms of k the fundamental pressure equation of kinetic theory *3.22* becomes

$$p = nkT \qquad (3.29)$$

Values of these various quantities to three significant figures are

$$N_A = 6.02 \times 10^{23}\ \text{mol}^{-1}$$
$$m_u = 1.66 \times 10^{-27}\ \text{kg}$$
$$m_p = 1.67 \times 10^{-27}\ \text{kg}$$
$$R = 8.31\ \text{J K}^{-1}\ \text{mol}^{-1}$$
$$k = 1.38 \times 10^{-23}\ \text{J K}^{-1}$$

Using these values with equation *3.28* we can calculate a typical speed for a nitrogen molecule at normal temperatures. With $M_r = 28$ and taking $T = 290$ K, the *root mean square speed* is

$$c_{\text{r.m.s.}} = \sqrt{\langle c^2 \rangle} = (3kT/m)^{\frac{1}{2}}$$
$$= (3kT/M_r m_u)^{\frac{1}{2}} \approx 508\ \text{m s}^{-1}$$

The formula for $c_{\text{r.m.s.}}$ may also be expressed in molar quantities. Multiplying top and bottom in the bracket by N_A, we get

$$c_{\text{r.m.s.}} = (3RT/M)^{\frac{1}{2}}$$

Comparing with formula *3.21* for the speed of sound, we find

$$c_{\text{r.m.s.}}/v_s = (3/\gamma)^{\frac{1}{2}}$$

so $c_{\text{r.m.s.}}$ is a little larger than the speed of sound. For air with $\gamma = 7/5$ (section 3.5.4) the ratio is 1.46.

We should note at this stage that $c_{\text{r.m.s.}}$ is not equal to $\langle c \rangle$: if you take the square root of the average of the squares of the molecular speeds, the result has the dimensions of speed, but is not equal to the average of the speeds themselves. We shall show how $c_{\text{r.m.s.}}$ and $\langle c \rangle$ are related in the Appendix, section A2. The result derived there is

$$\langle c \rangle = \sqrt{\frac{8}{3\pi}}\, c_{\text{r.m.s.}} = 0.921\, c_{\text{r.m.s.}} \qquad (3.30)$$

Since they only differ by about 8% it is often good enough in rough calculations to take $\langle c \rangle \approx c_{\text{r.m.s.}}$, the formula for which is easy to remember.

3.5 EQUIPARTITION OF ENERGY

According to equation *3.28*

$$\tfrac{1}{3}m\langle c^2 \rangle = kT$$

The left hand side is 2/3 of the mean kinetic energy of the molecules; the right hand side is proportional to temperature. It therefore turns out that *temperature is a measure of the energy of thermal motion*: mean kinetic energy is proportional to thermodynamic temperature. Can we be a little more precise? The

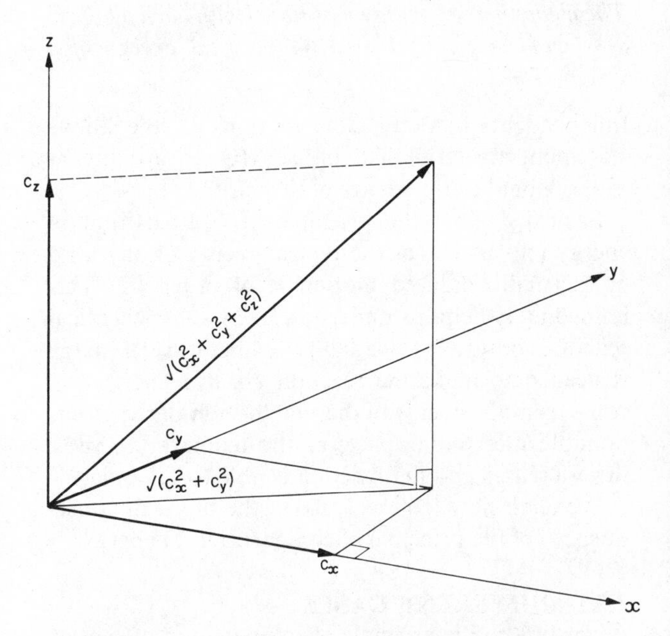

Figure 3.8 Vectorial addition of components of velocity

total velocity **c** is made up of three independent components which add vectorially (figure 3.8):

$$c^2 = c_x^2 + c_y^2 + c_z^2$$

where c_x, c_y and c_z are the components of the velocity in the x, y and z directions. The three terms are independent since they correspond to three different degrees of freedom*, x, y and z. We may therefore average each separately:

$$\langle c^2 \rangle = \langle c_x^2 \rangle + \langle c_y^2 \rangle + \langle c_z^2 \rangle$$

* 'Degree of freedom' is used here in just the same sense as when we describe the thermodynamic state of a system (page 2). There will be a minimum number of variables which must be used to specify the position of the system, which here is a molecule.

As there is no preferential direction for molecular motion, we must have

$$\langle c_x^2 \rangle = \langle c_y^2 \rangle = \langle c_z^2 \rangle = \tfrac{1}{3}\langle c^2 \rangle$$

Then the mean kinetic energy associated with *each* degree of freedom is

$$\tfrac{1}{2}m\langle c_x^2 \rangle = \tfrac{1}{6}m\langle c^2 \rangle = \tfrac{1}{2}kT$$

This is a special case of a very important result of classical physics which is stated in the *principle of equipartition of energy*:

The mean thermal energy associated with each independent quadratic contribution to the total energy of a system is $\tfrac{1}{2}kT$.

Independent quadratic contributions in the above statement are terms like $\tfrac{1}{2}m\langle c_x^2 \rangle$. (Quadratic means proportional to the square of the variable.)

Basically, what the principle of equipartition of energy tells us is that the typical energy of any kind of thermally induced motion is of order kT. This immediately helps us understand why some chemical reactions only take place at high temperature: if energy is needed to make the reaction go, this energy can come from the energy of thermal motion of the atoms or molecules taking part; as the temperature rises, this increases and the reaction is able to take place.

We shall now look at some of the physical consequences of the principle of equipartition of energy.

3.5.1 DIFFERENT GASES

According to the principle of equipartition of energy, different gases at the same temperature will have the same values for $\tfrac{1}{3}m\langle c^2 \rangle$, namely kT. Hence from equation *3.29*, $p = nkT$, we see that different gases at the same temperature and pressure contain the same number of molecules per unit volume. This observation is equivalent to *Avogadro's principle*:

Equal volumes of gases at the same pressure and temperature contain equal numbers of molecules.

Avogadro deduced this hypothesis from the fact that volumes of gases (measured at the same temperature and pressure) which react together chemically are in the ratios of small whole numbers. This suggested to him a direct relation between volume and number of molecules. This principle is only true for ideal gases.

The converse of Avogadro's principle must also be true; namely, that for all (ideal) gases at the same pressure and temperature, equal numbers of molecules occupy the same volume. The *molar volume at s.t.p.* of an ideal gas* is therefore a universal constant. Its value is

$$2.24 \times 10^{-2}\ \mathrm{m}^3$$

Again, according to equipartition, the molecules of all gases at the same temperature must on average have the same translational† kinetic energy. Heavier molecules will therefore move more slowly. This must be true, not only for normal gas molecules, but also for the fat globules or smoke particles in which we see Brownian movement. It is only because they are so heavy (compared with a normal gas molecule) that they move slowly enough for us to be able to watch their motion under a microscope.

3.5.2 DIFFUSION

Gases can pass slowly through porous materials like unglazed porcelain. Such materials contain networks of fine holes and channels through which the molecules move, but only with difficulty because the pores are so small that the molecules are continually colliding with the walls. The process is called diffusion. Now we might expect that the rate at which a molecule will diffuse will be proportional to its speed: the molecules which rattle around rapidly will progress quicker than those that move slowly between collisions with the walls. According to equation *3.28*, at a given temperature, molecular speed is inversely proportional to the square root of molecular mass. It follows from Avogadro's principle that the densities of different gases (measured at the same temperature and pressure) are proportional to the molecular masses. We would therefore conclude that *the rates of diffusion of different gases under the same conditions of temperature and pressure are inversely proportional to the square roots of their densities.*

This is *Graham's law of diffusion* which is borne out by experiment.

3.5.3 MIXTURES OF IDEAL GASES

Suppose we have a mixture of ideal gases. According to the simplifying assumptions we made when we set up the model of an ideal gas (page 45) the molecules

* s.t.p. = standard temperature and pressure, 273.15 K and 1 atm (1.01×10^5 Pa).

† *Translational* means associated with their movement in space. $\tfrac{1}{2}m\langle c^2 \rangle$ is translational kinetic energy. Later in this chapter we will consider other forms of energy which a molecule may have.

of each will move independently of the others so that the total pressure is given by adding together the pressures from each gas:

$$p = \tfrac{1}{3}(n_1 m_1 \langle c_1^2 \rangle + n_2 m_2 \langle c_2^2 \rangle + \ldots)$$

According to the principle of equipartition of energy, each of the terms $m_1\langle c_1^2 \rangle$ etc. has the value $3kT$. Hence the total pressure is

$$p = (n_1 + n_2 + \ldots)kT \qquad (3.31)$$

But $n_1 kT$ is the pressure gas number 1 would exert if it were present alone, and so with each of the others. Therefore,

The total pressure exerted by a mixture of ideal gases is equal to the sum of the pressures which would be exerted by each gas separately if the others were not present.

This is *Dalton's law*. Equation *3.31* may also be written

$$p = p_1 + p_2 + \ldots + p_i + \ldots$$

where $p_i = n_i kT$ is called the *partial pressure* of the ith gas. Dalton's law is only true for ideal gases. Figure 3.9 shows a simple experimental arrangement for verifying Dalton's law.

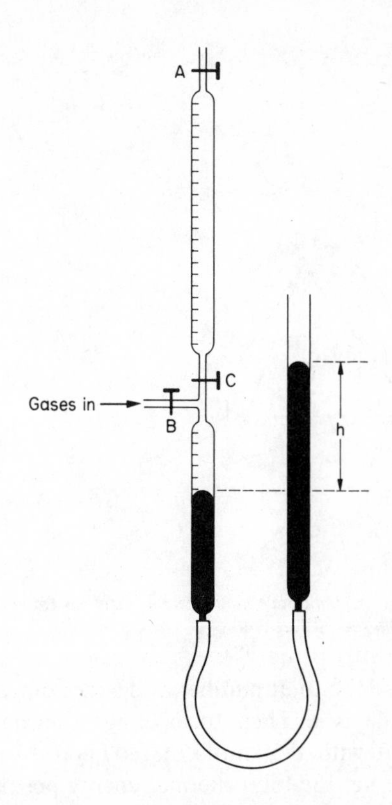

Figure 3.9 A simple arrangement for verifying Dalton's law. Gas is first forced out of both graduated cylinders by opening taps *A* and *C* and raising the mercury. *A* is then closed, and the mercury lowered to leave an (essentially) evacuated space in the upper graduated tube. Tap *C* is then closed. Known volumes of gases, admitted through *B*, are measured out in the lower cylinder at pressures set by the mercury level. They are then transferred to the upper cylinder by opening *C* and adjusting the mercury level.

3.5.4 HEAT CAPACITIES OF GASES

We have applied equipartition to the degrees of freedom corresponding to the motion of the molecule as a whole. There were three, one for each of the co-ordinates x, y and z which fix the molecule's position. Because they correspond to movement of the whole molecule in space, they are known as *translational degrees of freedom*. With polyatomic molecules (ones containing more than one atom) there will also be *internal degrees of freedom* corresponding to changes in the relative positions of the atoms within the molecule. A diatomic molecule (two atoms) has a total of five degrees of freedom: three translational corresponding to the position of the centre of mass, and two rotational corresponding to the two angles which are needed to fix which way the molecule points (figure 3.10). In a molecule with more than two atoms, there are generally three rotational degrees of freedom (figure 3.11). Each of these will give a quadratic contribution to the total energy: rotational kinetic energy of the form $\frac{1}{2}I\omega^2$ where I is moment of inertia and ω angular velocity. They will therefore also be subject to equipartition.

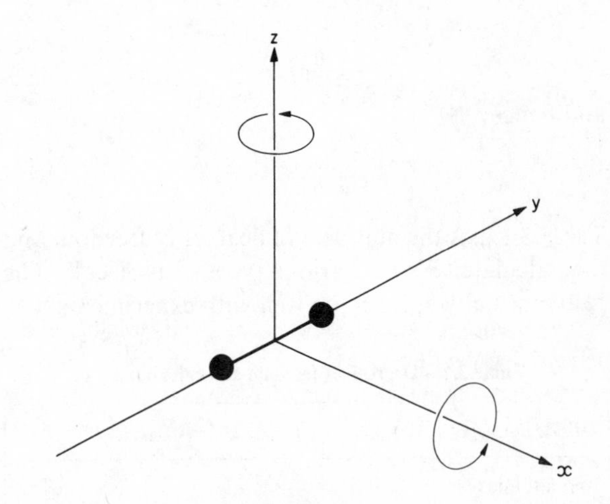

Figure 3.10 A diatomic molecule has two independent rotational degrees of freedom

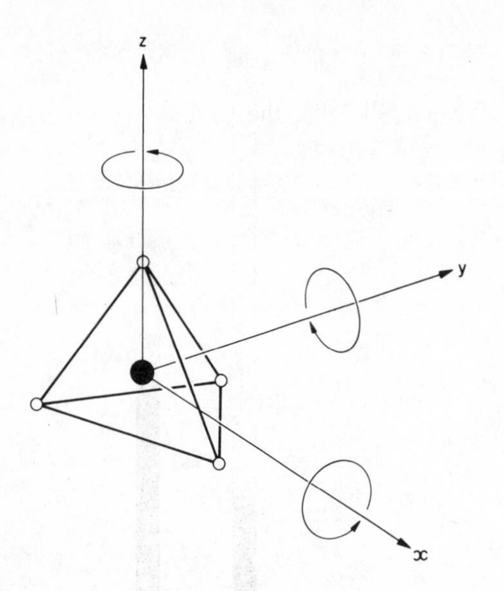

Figure 3.11 A polyatomic molecule generally has three independent rotational degrees of freedom

Suppose the total number of degrees of freedom of a molecule is f. Then the average thermal energy associated with that molecule is $f\frac{1}{2}kT$. Multiplying by N_A we get the total thermal energy per mole:

$$U_{\text{thermal}} = \tfrac{1}{2}fRT$$

Hence, for n mol,

$$C_V = \left(\frac{\partial Q}{\partial T}\right)_V = \left(\frac{\partial U}{\partial T}\right)_V = \tfrac{1}{2}fnR \qquad (3.32)$$

But, by equation 3.8,

$$C_p - C_V = nR$$

so

$$C_p = \frac{n}{2}(f + 2)R \qquad (3.33)$$

and hence

$$\gamma = \frac{C_p}{C_V} = (f + 2)/f \qquad (3.34)$$

Table 3.1 lists the numbers of degrees of freedom and the calculated γ's for various types of molecule. The results usually agree quite well with experiments.

Table 3.1 Degrees of freedom and calculated γ's

Molecular type	$f_{rotation}$	$f_{translation}$	f_{total}	γ
monatomic	0	3	3	5/3
diatomic	2	3	5	7/5
polyatomic	3	3	6	4/3

3.5.5 OSCILLATORS

Simple harmonic oscillators (of which a pendulum is a simple mechanical example) also store energy proportionally to the square of a coordinate. The total energy of an undamped simple harmonic oscillator is constant, but it flows backwards and forwards between potential and kinetic forms. In a pendulum, for example, the energy is entirely in kinetic form in the middle of the swing while it is entirely in potential form at the extremes of the motion where the velocity is zero. Both these forms are expressed as the square of a coordinate: kinetic, $\frac{1}{2}m\dot{x}^2$, potential $\frac{1}{2}\alpha x^2$ where x is displacement, m mass, and α the restoring force constant. Now an oscillator oscillating at large amplitude is not in thermal equilibrium with its surroundings, and the amplitude will gradually decay as a result of viscosity, friction or other energy-loss mechanisms. It will never come totally to rest, however, because thermal motions will always ensure that there is some average residual energy. When only this thermal energy is left, the oscillator is in thermal equilibrium with its surroundings and equipartition will apply. Since both potential and kinetic energies will average $\frac{1}{2}kT$, the mean total energy will be kT. The total energy at any instant will not, of course, always be the same, but will fluctuate around this mean value.

Let us apply equipartition to a simple pendulum to find its root-mean-square amplitude of motion under thermal conditions. When deflected through a small angle θ, the potential energy increases by $\frac{1}{2}mgl\theta^2$ where m is the mass of the bob and l the length. Then, by equipartition,

$$\tfrac{1}{2}mgl\langle\theta^2\rangle = \tfrac{1}{2}kT$$

and

$$\theta_{\text{r.m.s.}} = (kT/mgl)^{\frac{1}{2}}$$

In electrical conductors, the random thermal motions of electrons produce small fluctuating currents and voltages. These thermally produced signals are known as *Johnson noise*. ('Noise' because, if you amplify them and listen to them, the sound is a featureless noise.) Again each mode of motion of current in the circuit is subject to equipartition. General analysis is complicated, but there are simple aspects we can describe. Oscillatory circuits can be analyzed in the same way as mechanical oscillators, but here the equivalent of potential energy is energy stored on the capacitor (page 22): $Q^2/2C$, and the kinetic form corresponds to the current (charge in motion) in the inductor (page 22): $LI^2/2$. Analysis of Johnson noise in a resistor is complex; but, again, we might expect the thermally produced voltages to have a mean square value proportional to thermal energy, i.e. proportional to T, giving $V_{r.m.s.} \propto \sqrt{T}$. The proper analysis shows that this guess is correct. Johnson noise is important in sensitive electrical apparatus because it obviously sets a limit to the smallness of signals which can be measured: one cannot hope to detect signals which are much smaller than the noise produced in the electrical circuits of the detector itself. This limitation is crucial in, for example, radio astronomy where very weak signals from the extremes of the known universe have to be measured, or in the reception of signals from distant space vehicles.

3.5.6 SOLIDS

Although the *mean* positions of the atoms in a solid are fixed with respect to one another, the atoms are free to vibrate about the mean positions within the small space available to them. Each atom has three independent degrees of freedom corresponding to the three mutually perpendicular directions in which it can move. Hence, each atom should behave like three simple harmonic oscillators and, by equipartition, should have $3 \times kT$ of thermal energy on the average. The molar thermal energy of a pure element (N_A atoms) should then be $3RT$ and the molar heat capacity, $3R = 24.9\ \mathrm{J\,K^{-1}\,mol^{-1}}$. This is *Dulong and Petit's law* which is obeyed quite well at normal temperatures in most cases.

3.5.7 BREAKDOWN OF EQUIPARTITION

Dulong and Petit's law fails at low temperatures because the principle of equipartition fails. The principle of equipartition is a *classical* result and therefore takes no account of *quantization.* According to quantum theory, energy comes in indivisible packets called *quanta.* The size of the packet depends on the nature of the system. In an electromagnetic wave, for example, the quanta are called *photons* and they have energy $h\nu$ where h is the Planck constant and ν the frequency of the wave.* In atoms, quantization leads to the various atomic energy levels which are responsible for line spectra. According to quantum theory, energy is *always* quantized, and this must also apply to the thermal energy of a solid. Now as long as the permitted energy steps are much smaller than the typical thermal energy, kT, they are not noticed: the energy structure is too fine-grained to be seen. But if kT is smaller than the energy steps, equipartition cannot work because, to satisfy equipartition the energy would have to take up a value *between* steps and that is forbidden by quantum theory. This is what happens in solids at low temperatures: kT becomes too small to make the necessary steps in energy; less thermal energy gets stored than equipartition would predict and heat capacities fall (figure 3.12).

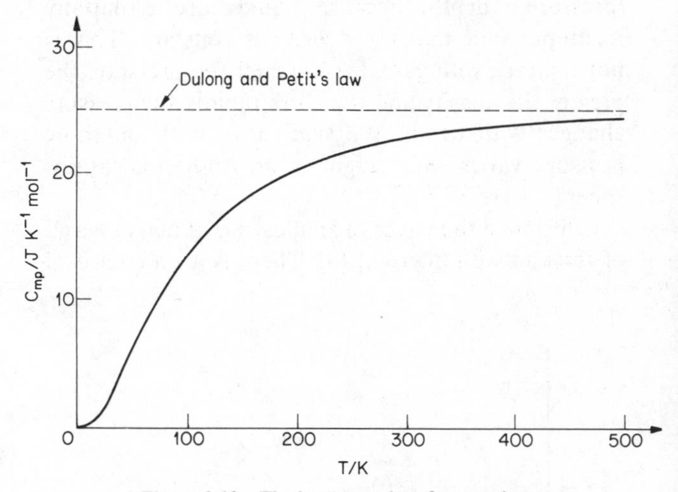

Figure 3.12 The heat capacity of germanium

When we discussed the degrees of freedom of diatomic gases to calculate the principal heat capacities and γ, we only included the three translational motions and *two* rotations. We did not include spin about the common axis of the atoms (figure 3.13). The reason for this is that a simple calculation based on the ideas of quantum theory shows that the energy

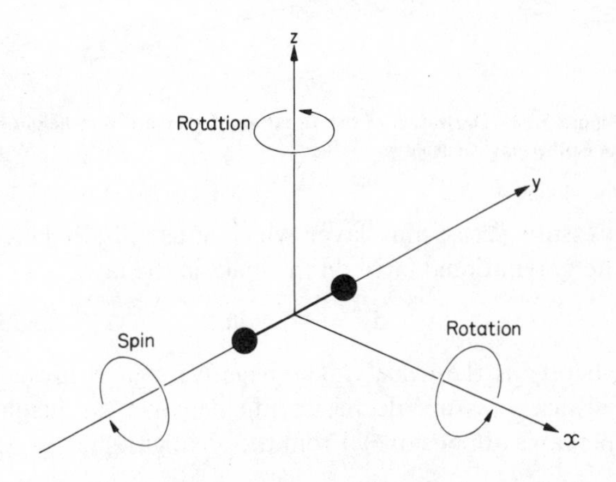

Figure 3.13 Rotation and spin of a diatomic molecule

* ν is the Greek letter *nu*.

of a single quantum of spin is much greater than kT at normal temperatures. Hence this motion is 'frozen out': the temperature is so low that thermal energies are not sufficient to excite it.

3.6 ISOTHERMAL GAS IN A GRAVITATIONAL FIELD

In a liquid, pressure increases linearly with depth (pressure $\propto$ depth) because liquids are essentially incompressible and the density is constant. This is not the case with gases: the greater the pressure, the greater the density and the more rapidly the pressure changes with depth. We shall now work out how pressure varies with height in an isothermal atmosphere.

Consider a thin layer of an ideal gas at height h and of thickness $\mathrm{d}h$ (figure 3.14). There is a difference of pressure across this layer which must just balance the gravitational force on the mass in the layer:

$$\mathrm{d}p = -\rho g\,\mathrm{d}h \qquad (3.35)$$

where ρ is the density. The negative sign is present because pressure decreases ($\mathrm{d}p$ negative) as height increases ($\mathrm{d}h$ positive). From the ideal gas law,

$$\rho = M/V_m = Mp/RT$$

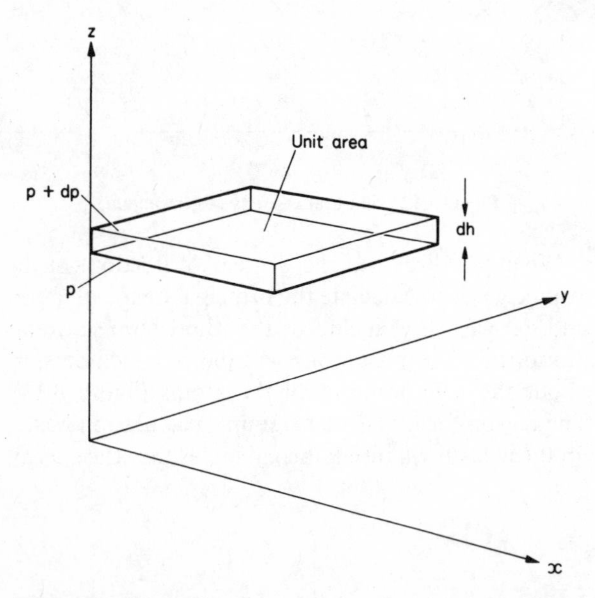

Figure 3.14 Derivation of the variation of pressure with height in an isothermal atmosphere

Substituting for ρ in *3.35* gives the differential equation

$$\frac{\mathrm{d}p}{p} = -\frac{Mg}{RT}\mathrm{d}h$$

Putting in the limits of integration,

$$\int_{p_0}^{p}\frac{\mathrm{d}p}{p} = -\frac{Mg}{RT}\int_0^h \mathrm{d}h$$

where $p = p_0$ at $h = 0$ (say ground level) and p is the pressure at height h. This gives, on integration,

$$\ln(p/p_0) = -Mgh/RT$$

Putting as an exponential,

$$p = p_0 \exp(-Mgh/RT) = p_0 \exp(-h/H) \qquad (3.36)$$

We find that pressure decreases exponentially with height. $H = RT/Mg$ is called the *scale height*. It is the height in which the pressure decreases by e^{-1}, that is by 1/2.718. Figure 3.15 shows p/p_0 plotted

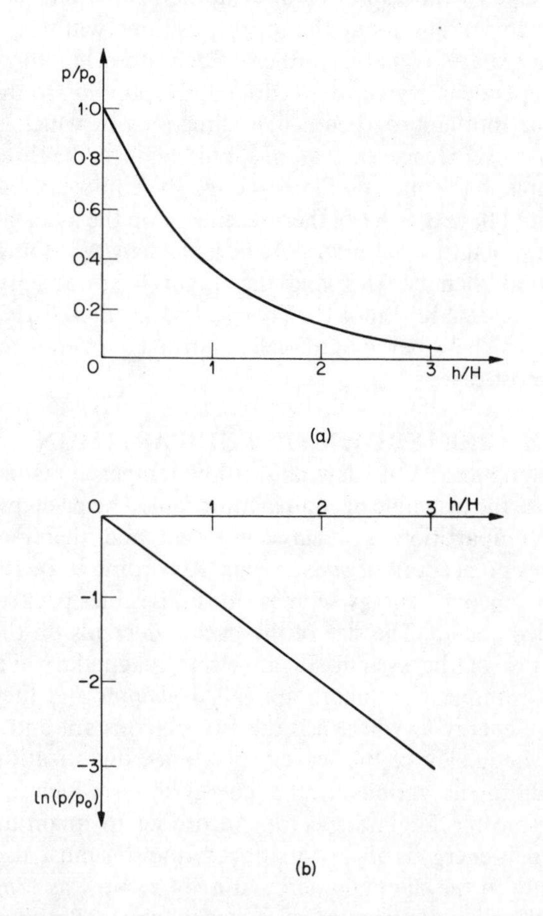

Figure 3.15 The dependence of pressure on height in an isothermal atmosphere. The pressure scale is linear in (a) and logarithmic in (b).

against h/H. Since we are assuming an isothermal atmosphere, density is proportional to pressure, so density decreases with height in the same way. We therefore also have for density ρ and number of molecules per unit volume n

$$\rho = \rho_0 \exp(-h/H) \tag{3.37}$$

$$n = n_0 \exp(-h/H) \tag{3.38}$$

where ρ_0 and n_0 are the density and number density at height $h = 0$.

Substituting for the earth's atmosphere we find $H \approx 8$ km. This gives the pressure at the top of Everest (8.88 km) about $\exp(-1.11) \approx 1/3$ of that at the ground. The earth's atmosphere is not, of course, isothermal; temperature generally decreases with height; so this law is not obeyed accurately.

3.6.1 BOLTZMANN FACTORS

If we substitute for H in *3.38* in terms of molecular quantities we have

$$n = n_0 \exp(-mgh/kT) = n_0 \exp(-\phi/kT)$$

where $\phi = mgh$ is the potential energy of a molecule at height h. This is a special case of a general and very important result of statistical mechanics which says that

the probability of finding a system in a state of energy ε is proportional to $\exp(-\varepsilon/kT)$.

This result applies whatever the form of the energy (kinetic or potential). The exponential expression is known as a *Boltzmann factor*.

To apply this in a simple case, suppose that an atom can be in one of two quantum states with energies ε_1 and ε_2. The probability that it is in state 1 will be

$$P_1 = A \exp(-\varepsilon_1/kT)$$

where A is the constant of proportionality; and the probability that it is in state 2 will be

$$P_2 = A \exp(-\varepsilon_2/kT)$$

The ratio of probabilities is therefore

$$P_1/P_2 = \exp(-\varepsilon_1/kT)/\exp(-\varepsilon_2/kT) = \exp(-\Delta\varepsilon/kT)$$

where $\Delta\varepsilon$ is the energy difference between the states. If $\Delta\varepsilon \gg kT$, the chance of being excited out of the lower state by thermal energies is negligible. If the levels are separated by* $1\ \text{eV} = 1.6 \times 10^{-19}$ J, a typical atomic energy level spacing, the value of the Boltzmann factor at room temperature is

$$\exp(-1.6 \times 10^{-19}/1.38 \times 10^{-23} \times 290)$$
$$= \exp(-40) \approx 4 \times 10^{-18}$$

so if we observe the atom a million million million times, we shall only find it excited thermally into the upper state four times! On the surface of the sun, where the temperature is about 6000 K, the Boltzmann factor would be $\exp(-1.93) \approx 1/6.9$ so atoms are often excited thermally and the sun shines.

3.7 DISTRIBUTION OF MOLECULAR SPEEDS

In the kinetic theory derivation of pressure in an ideal gas, we came out with a result involving the mean square molecular speed. The molecules will not, of course, all be moving at the same speed, nor will the speed of any one molecule be constant. A molecule's speed will generally be changed whenever it makes a collision with another molecule or with the walls of the container. However, we would expect that when a gas is in equilibrium, the molecular motions would settle down so that, on average, the number of molecules moving in any particular range of speeds would always be the same. If, for example, we did a series of measurements to find how many had speeds between 500 and 510 m s^{-1} we would expect always to get about the same answer. In other words, we would expect the molecules to be distributed over the range of possible speeds in some characteristic way. Such a distribution is described by a *distribution function* $f(c)$ which is defined:

$f(c)\,\mathrm{d}c$ is the fraction of the molecules moving with speeds between c and $(c + \mathrm{d}c)$.

Thus, $f(500)$ is the fraction of all molecules whose speeds lie between 500 and 501 m s^{-1}; $f(500)/100$ is the fraction with speeds between 500 and 500.01 m s^{-1}; and so on.

* The *electronvolt*, symbol eV, is a conveniently sized unit for discussing microscopic energies. One electronvolt is the energy an electron gains in falling through a potential difference of one volt. Since the charge on an electron is -1.6×10^{-19} C, this energy is, by definition of the volt, 1.6×10^{-19} J.

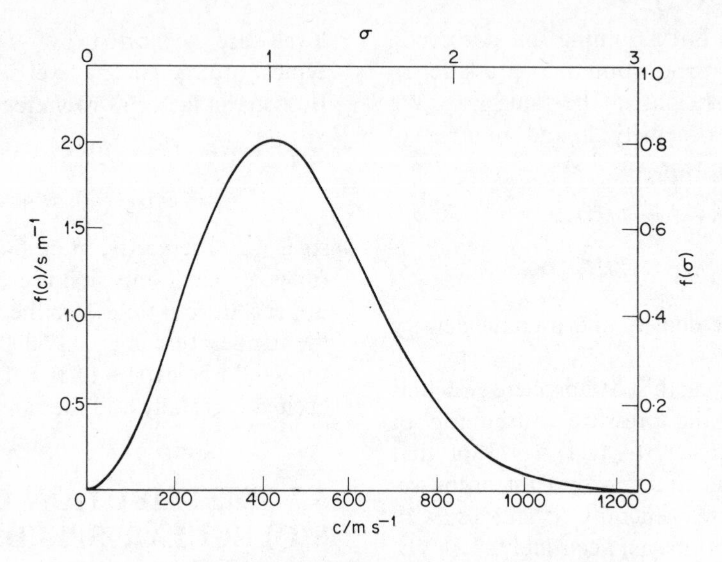

Figure 3.16 The Maxwell distribution of molecular speeds. The calibrations on the lower and left axes are for nitrogen at 290 K. The calibrations on the upper and right axes are in dimensionless form and so represent a universal curve from which values can be read off for any gas. The dimensionless parameter is $\sigma = (mc^2/2kT)^{\frac{1}{2}}$.

The form of the distribution function for molecular speeds in a gas is

$$f(c) \propto c^2 \exp(-mc^2/2kT) \qquad (3.39)$$

This result, known as the *Maxwell distribution* of molecular speeds, can be derived using the theory of statistical mechanics and has been verified experimentally.

The exponential term in *3.39* looks like a Boltzmann factor (see last section), and indeed it is. It has the form exp (−kinetic energy/kT). But why is there a c^2 in front? The reason is that the faster the speed, the more possible ways there are of choosing the velocity of the molecule. The number of ways increases as the square of the speed. (To prove this requires statistical mechanics.) Thus, the c^2 in front is proportional to the number of states of motion available and the exponential is proportional to the chance that a molecule is in any one of those states.

The Maxwell distribution is illustrated in figure 3.16. The calibrations on the left and lower axes are for nitrogen at 290 K. The calibrations on the upper and right axes are in *dimensionless* form and so represent a *universal* curve from which one can read off values that can be applied to any gas. The dimensionless parameter is $\sigma = (mc^2/2kT)^{\frac{1}{2}}$.

Figure 3.17 shows a suitable experimental arrangement for measuring the distribution of molecular speeds. Vapour is produced in an oven and escapes through a small hole in one side as a molecular beam. The surrounding space is kept at high vacuum so as to prevent the molecules in the beam from being scattered by collisions with other molecules. The beam is directed parallel to the axis of a simple velocity selector which consists of two discs with slits mounted on a common axle which rotates in the direction shown. As the first slit crosses the beam, a short burst of molecules is allowed through; but these are moving with a variety of speeds. Most will strike the second disc. Only those whose speed is such that the second slit is in line with the beam by the time they arrive at the second disc will pass through and reach the detector. If the distance between the discs is L, the angular velocity* ω and the angle between the slits θ, the speed selected is $L\omega/\theta$. By measuring the number reaching the detector as ω is varied, the speed distribution can be plotted out.†

* ω is the Greek letter *omega*.

† Although the speed distribution of the molecules inside the oven is Maxwellian, that of the molecules coming out in the beam is not, because the faster ones have more chance of arriving at the hole and escaping. It turns out that the resulting speed distribution in the beam goes as $c^3 \exp(-mc^2/2kT)$.

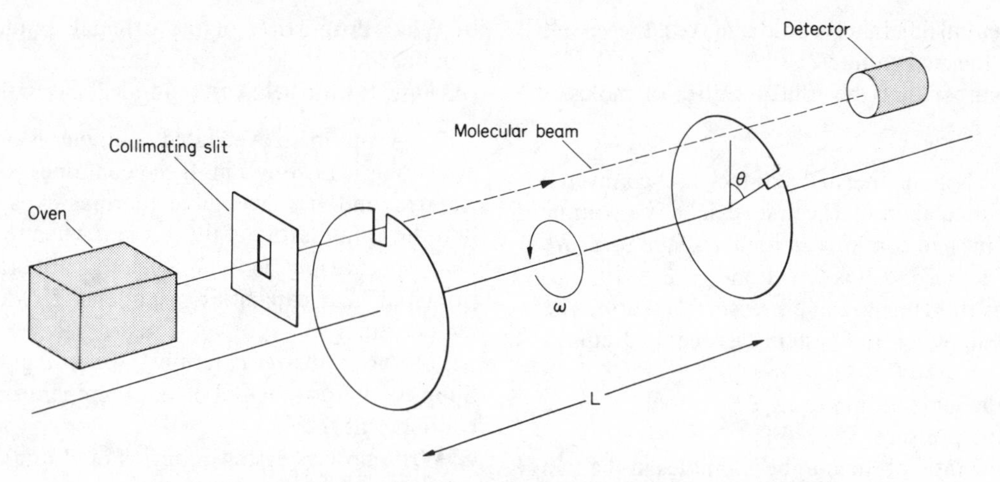

Figure 3.17 An arrangement for measuring the distribution of molecular speeds

PROBLEMS

Ideal gas: Boyle's law, equation of state; Joule's law, internal energy

3.1 Use equation *3.3* with the known value of R to calculate the molar volume of nitrogen at 10 °C and 1 atm pressure.

3.2 A mercury barometer has a little air in the space above the mercury. The height of the mercury column is 758 mm and the length of tube above the column is 80 mm. When the tube is lowered a little the height of the mercury column is 755 mm and the length of the space above is 50 mm. What is atmospheric pressure?

3.3 A toy hot-air balloon consists of a spherical envelope 1.5 m in diameter below which is suspended a burner. Air heated by the burner enters the balloon through an opening at the bottom and collects in the envelope. If the total mass is 0.3 kg, how hot does the air in the envelope become before the balloon rises?
(Ignore any change in composition of the air due to products of combustion and take the ambient temperature as 15 °C.)
[Density of air at s.t.p. = 1.29 kg m^{-3}.]

3.4 Two vessels of volume 1×10^{-3} m^3 and 3×10^{-3} m^3 are connected by a capillary tube and contain air. Initially they are at the same temperature, 15 °C, and the pressure is 1 atm. The larger vessel is then heated to 100 °C while the smaller remains at 15 °C. What is the final pressure?
[*Hint*: remember to conserve the total amount of gas present.]

3.5 A bubble of radius r is blown from a soap solution of surface tension γ.
a) Show that the work done on the surface is $8\pi\gamma r^2$.
The gas has to be compressed slightly above the pressure of the surrounding atmosphere in order to force it into the bubble.
b) Show that the work done in compressing an ideal gas isothermally by a small amount dp is V dp.
c) Hence show that the work done on the gas in blowing the bubble (isothermally) is $\frac{16}{3}\pi\gamma r^2$.
d) Hence, find the work required to blow a soap bubble of radius 100 mm from a soap solution of surface tension 2.5×10^{-2} N m^{-1}.

3.6 Explain how Joule's law follows from the fact that an ideal gas does not change its temperature when it makes a free expansion.

Show how it follows from Joule's law that changes in the internal energy of an ideal gas may always be written $\mathrm{d}U = C_V\,\mathrm{d}T$.

3.7 Two vessels in thermal contact are connected by a tube with a tap in it. The first vessel has a volume of 2×10^{-3} m^3 and contains an ideal gas at a pressure of 2 atm. The second vessel has a volume of 3×10^{-3} m^3 and is evacuated. The tap is opened. When the system has reached equilibrium,
a) what is the temperature change?
b) what is the pressure?
c) what is the ratio of the masses of gas in the two vessels?

d) what essential difference would it make if the vessels were not in thermal contact?
[*Hint*: remember that the total number of moles is constant.]

3.8 Two vessels in thermal contact are connected by a tube with a tap in it. The first vessel has a volume of 2×10^{-3} m^3 and contains an ideal gas at a pressure of 2 atm. The second has a volume of 3×10^{-3} m^3 and contains the same gas at a pressure of 4 atm. The tap is opened. When the system has reached equilibrium,
a) what is the temperature change?
b) what is the pressure?
c) what is the ratio of the number of moles in the two vessels?

[*Hint*: remember that the total number of moles of gas is constant. Write down the ideal gas equation for each vessel in its initial and final states with the numbers of moles appearing in the equations as unknowns. Then eliminate the *n*'s between the equations.]

Adiabatic changes in an ideal gas; difference of principal heat capacities; adiabatic equations

3.9 Why, on the average, does temperature decrease with height in the atmosphere?

3.10 A bicycle pump compresses air from 1 atm to 3 atm before the valve opens and the air enters the tyre. If the ambient temperature is 290 K, how hot can the end of the pump become?
[For air, $\gamma = 1.4$.]

3.11 A diesel engine takes in atmospheric air and compresses it adiabatically so that it becomes hot. Fuel is then injected into the compressed air and the fuel–air mixture ignites provided that the temperature is high enough. Work is obtained as the resulting high pressure gases force back the piston. If the ignition temperature of diesel fuel is 430 °C and the air enters the engine at 17 °C, what is the minimum compression ratio (ratio of initial volume to compressed volume) which must be used?
[For air, $\gamma = 1.4$.]

3.12 A cylinder contains compressed air at 50 atm and 290 K. The tap is opened to the atmosphere.
a) What is the temperature of the air left in the cylinder when the pressure has dropped to 1 atm assuming that the process is adiabatic?
b) What proportion of the original amount of gas remains?
[Assume the air behaves as an ideal gas with $\gamma = 1.4$.]

3.13 A tap on an evacuated container is opened and atmospheric air flows in. If the container is thermally isolated and has negligible thermal capacity, show that the temperature of the gas in it when the pressure reaches atmospheric is γT_0 where γ is the ratio of the principal heat capacities of air and T_0 atmospheric temperature.
Set out your answer carefully, proceeding as follows: Suppose there are n mol of air at temperature T in the container at the end.
a) Write an expression in terms of R and T_0 for the work done by the atmosphere in pushing the n mol into the container.
b) By how much is the internal energy of the n mol greater at the end than when they were outside the container before the tap was opened?
c) Assuming the air to approximate to a perfect gas, write another expression for the change in internal energy in terms of the rise in temperature and a heat capacity.
d) Hence find T. (Remember equation *3.9*!)

3.14 A vessel with a tap contains a gas at a pressure of 1000 mmHg. The tap is opened so that the pressure drops to atmospheric (760 mmHg) and is quickly closed again. After the gas remaining in the vessel has returned to the temperature of the surroundings, the pressure is found to be 848 mmHg. What can you deduce from these observations?

3.15 A bicycle pump is used to pump up a tyre in which the (total) pressure is 3 atm. The volume of atmospheric air compressed in each stroke of the pump is 100 ml and the ambient temperature is 290 K. Assuming that:
i) the compression is adiabatic;
ii) the valve in the tyre opens when the pressures in the pump and tyre are equal;
iii) the volume in the tyre is large compared to the volume in the pump;
iv) the volume remaining in the pump when the handle is fully home is negligible;
a) sketch a graph showing how the pressure in the pump varies as the pump handle is pushed in.
Calculate
b) the volume to which the gas has to be compressed before the pressure in the pump equals that in the tyre,

c) the work done in compressing the gas to the pressure at which the valve opens,
d) the work done in forcing the gas into the tyre, and
e) the total work done in one stroke of the pump.
[For air, $\gamma = 1.4$.]

3.16 A compressor takes in air at pressure p_1 and temperature T_1 and delivers compressed air at pressure p_2 at a rate (volume per second) V. If the compression is adiabatic,
a) show that the volume per second taken in by the compressor is $V(p_2/p_1)^{1/\gamma}$,
b) show that the number of moles the compressor takes in per second is $p_1^{(\gamma-1)/\gamma}p_2^{1/\gamma}V/RT$,
c) find the temperature T_2 of the air after compression,
d) show that the work required to compress 1 mol is $\gamma R(T_2 - T_1)/(\gamma - 1)$ (see section 2.5),
e) show that the power required to run the compressor is

$$\frac{\gamma}{\gamma - 1} p_1 V \left(\frac{p_2}{p_1}\right)^{1/\gamma} \left(\frac{T_2}{T_1} - 1\right)$$

A real compressor has an overall efficiency of 65%. It takes in atmospheric air at 300 K and delivers compressed air at 10 atm at a rate of 10^{-4} m^3 s^{-1}.
f) How many moles pass through the compressor per second?
g) At what temperature is the compressed air delivered?
h) What is the power required to run the compressor?
[For air, $\gamma = 1.4$.]

Bulk moduli and compressibilities of an ideal gas; speed of sound

3.17 Define bulk modulus. Explain why the adiabatic bulk modulus of a gas is larger than the isothermal bulk modulus. Show that their ratio is γ.

3.18 The frequency of an organ pipe when blown with dry air is 440 Hz. When blown with argon it drops to 413 Hz. Explain why the pitch changes. What is the ratio of the speed of sound in argon to that in air? Deduce the value of γ for argon given that $\gamma_{air} = 1.4$ and that at the temperature of the experiment $\rho_{air} = 1.21$ kg m^{-3} and $\rho_A = 1.67$ kg m^{-3}.

3.19 In an experiment to measure the speed of sound in hydrogen, sound waves of frequency 10 kHz are propagated down a tube filled with the gas. An oscilloscope whose horizontal scan is synchronized to the oscillator driving the sound source shows the signal picked up by a small microphone in the tube. As the microphone is moved along the tube away from the source, the positions are noted at which the waveform displayed on the oscilloscope passes through the same position on the screen:

distance moved/mm: 129 260 390 519 650

a) What is the speed of sound in hydrogen?
b) Given that the molar mass of hydrogen is 2.0 g and the temperature of the experiment 18 °C, what is γ for hydrogen?

Elementary kinetic theory: basic and simplifying assumptions; expressions for pressure; the mole, N_A, M_r, A_r, m_u, M, k; molecular speeds

3.20 Give an account of the essential features of the kinetic theory model of an ideal gas. Show how the model leads to the identification of thermodynamic temperature as a measure of molecular kinetic energy.

3.21 The density of air at s.t.p. is 1.29 kg m^{-3}. What is the molar mass?
[Use the value of R.]

3.22 Molecules of oxygen ($M_r = 32$) escape from the surface of the moon where the surface temperature is 50 °C.
a) What is their root mean square speed?
b) What is their mean kinetic energy?
c) What is their potential energy at the moon's surface?
d) Will their speed be sufficient for them to escape from the gravitational attraction of the moon?
e) Why does the moon have no atmosphere? (*Hint*: section 3.7.)
[Radius of moon = 1738 km. Acceleration of free fall at moon's surface = 1.62 m s^{-2}.]

3.23 What is the mass of oxygen in a high pressure cylinder, internally 750 mm long and 120 mm in diameter, which is filled to a pressure of 130 atm at 15 °C?
[Relative molecular mass of oxygen = 32.]

3.24 Approximately how many molecules are there in one litre of air?

3.25 For nitrogen, the molar volume at s.t.p. is 2.24×10^{-2} m^3 and the relative molecular mass is 28. What are
a) the mass of one molecule,
b) the mass of one mole,
c) the number of molecules in 10^3 mm^3 at s.t.p.,
d) the root mean square molecular speed at s.t.p.?

3.26 Given that 4 kg of helium at s.t.p. occupy 22.4 m^3, calculate the root mean square speed of helium atoms at 15 °C.

Equipartition of energy: Avogadro's principle; Graham's law of diffusion; mixtures of ideal gases, Dalton's law of partial pressures; heat capacities of gases, the value of γ; oscillators; heat capacity of solids, Dulong and Petit's law; breakdown of equipartition, quantization

3.27 Certain molecules will move on the surface of a liquid as a two-dimensional 'gas'. Analyze their motions using the ideas of kinetic theory and the principle of equipartition to show that they exert a force per unit length on the boundary of the surface equal to nkT where n is the number of molecules per area of surface.

3.28 State Avogadro's principle and show how it follows from the principle of equipartition of energy when it is applied to ideal gases.

3.29 Describe how the molecules of a gas diffuse through a porous material. How would you expect the rate of diffusion to depend on (a) temperature, and (b) molecular mass? Would the same ideas apply to the diffusion of a molecule through a gas or a liquid?

3.30 Explain the physical assumptions which must be made in order to derive Dalton's law of partial pressures.

A flask of volume 10^{-3} m^3 is initially evacuated. Volumes V of various gases are measured at pressures p and then transferred to the flask:

Gas:	O_2	N_2	H_2	He
$V/10^{-6}$ m^3:	15	17	28	50
p/mmHg:	15	8	30	45

What is the final pressure assuming the temperature is constant?

3.31 Atmospheric air contains 21% by volume of oxygen.
a) What is the partial pressure of oxygen?
b) What is the proportion of oxygen molecules?

3.32 Oxygen is collected at 18 °C over water. The measured volume at 765 mmHg is 8×10^{-5} m^3. Calculate
a) the volume of dry oxygen at s.t.p.
b) the mass of oxygen collected.
[Saturated vapour pressure of water at 18 °C = 15.5 mmHg. Relative molecular mass of oxygen = 32.]

3.33 Air is trapped in a uniform tube, closed at one end, by a drop of carbon tetrachloride. At 15 °C the length of the column of trapped air is 98 mm. At 30 °C it is 113 mm. If the saturated vapour pressure of carbon tetrachloride at 15 °C is 63 mmHg what is its saturated vapour pressure at 30 °C?
[Atmospheric pressure = 760 mmHg.]

3.34 A mixture of ideal gases contains n_A mols of gas A and n_B mols of gas B. Show that for the mixture $C_p - C_V = (n_A + n_B)R$.
[*Hint*: Follow the arguments leading to equation *3.8* but using Dalton's law to write the total pressure in terms of n_A and n_B.]

3.35 Explain why the value of the ratio of the principal heat capacities of gases depends on the number of degrees of freedom of its molecules. Show that $\gamma = (f + 2)/f$ where f is the total number of degrees of freedom. What values would you expect for γ for He, N_2, NH_3? What do you deduce from the fact that γ for mercury vapour is 5/3?

3.36 A 1 g mass hangs by a fine thread of length 1 m. Apply the principle of equipartition of energy to find the root mean square displacement of the mass due to thermal motion. Could such small displacements be measured?

3.37 A radio receiver is tuned by a resonant circuit consisting of an inductor and capacitor of capacitance C connected in parallel.
a) Show that the mean square charge on the capacitor due to thermal motions is $\langle Q^2 \rangle = kTC$.
b) What is the corresponding mean square voltage across the capacitor?
c) What is the r.m.s. voltage in the case where $C = 10^{-10}$ F and $T = 290$ K?
d) What is the significance of such thermal 'noise' in connection with the useful sensitivity of radio receivers?

3.38 Explain how the principle of equipartition leads to Dulong and Petit's law for the heat capacity of solids.

Why is Dulong and Petit's law not obeyed at low temperatures?

3.39 A tuning fork has a frequency of 440 Hz.
a) What is the size of one quantum of energy of vibration?
b) To approximately what temperature would the fork have to be cooled before equipartition would cease to apply to its thermal motions?

3.40 Absorption spectra of gaseous HCl show a line at 3.5 μm which results from molecular vibration (oscillation of the interatomic distance).
a) What is the frequency of vibration?
b) What is the energy of one quantum of vibrational energy?
c) To approximately what temperature would the gas have to be heated to excite vibrations thermally?

Isothermal gas in a gravitational field; Boltzmann factors

3.41 Why does pressure increase with depth linearly (i.e. directly proportional) in a lake, but exponentially in the atmosphere?

3.42 Show that pressure in an isothermal atmosphere decreases with height h as $\exp(-h/H)$ where H is a constant. Deduce a value for the scale height H of the atmosphere.

Jet aircraft on long distance flights typically fly at heights of about 10 km. What is the atmospheric pressure at this height?
[Take atmospheric density at ground level = 1.25 kg m^{-3}.]

3.43 What is a Boltzmann factor?

The wavelength of the sodium D lines is about 589 nm.
a) What is the frequency of the light?
b) What is the energy of one quantum of the light? (Give the answer in joules and electronvolts.)
c) The atom emits a quantum when it makes a transition from a state of higher energy to one of lower, the difference in the energy of the states being equal to the quantum energy carried off by the photon. What is the probability that an atom will be thermally excited into the upper state at 1000 K?

Distribution of molecular speeds, the Maxwell distribution

3.44 Why do the molecules of a gas not all move at the same speed? Explain how the probabilities of finding molecules with different speeds are described mathematically by a *distribution function.* Sketch the form of the Maxwell distribution.

3.45 In the apparatus shown in figure 3.17, $L = 0.5$ m, $\omega = 1000$ rad s^{-1}, $\theta = 1.0$ rad, the distance between the axis and the beam is 100 mm and the width of the slits 2 mm.
a) How do the fastest molecules selected get through?
b) How do the slowest molecules selected get through?
c) What range of speeds is selected?

3.46 If the moon had an atmosphere, molecules with sufficient speed would be able to escape from the gravitational attraction of the moon and be lost into space.
a) What is the escape speed for an oxygen molecule?
b) Make a rough estimate of the proportion of molecules in a lunar oxygen atmosphere at 300 K which would have enough speed to escape. (Since the exponential dominates the distribution at the speeds concerned, you may approximate the Maxwell distribution by $c \exp - (mc^2/2kT)$ so as to be able to evaluate the integral.)
c) Why is there no atmosphere on the moon?
[At the moon's surface, $g = 1.6$ m s^{-2}. Radius of the moon = 1700 km. Relative atomic mass of oxygen = 32.]

4 Real Substances

When developing the kinetic model for the ideal gas, we made two basic assumptions: that the size of the molecules was negligible, and that the attractive forces between them were negligible. Although these are reasonable approximations for gases at low density when molecules are very far apart, molecular size and intermolecular attraction cannot be ignored as the density increases. If there were no intermolecular attractive forces, there would be no condensed states of matter—no liquids or solids—because there would be nothing to hold the molecules together. Similarly, if molecules did not have a finite size, there would be nothing to prevent the collapse of matter in a condensed state to a point (of infinite density but finite mass). In this chapter, we shall examine some of the ways in which finite molecular size and intermolecular attractions determine the behaviour of matter in bulk. In the next chapter, we shall look at the nature of intermolecular forces in a little more detail and explore some of the consequences from a more microscopic point of view.

4.1 MEAN FREE PATH IN A GAS

If molecular size were truly negligible in a gas, there would be no intermolecular collisions and molecules would travel in straight lines between impacts with the walls of the container. This would have startling physical consequences. For example, thermal conductivity would be enormous, for a molecule which had collected extra thermal energy by colliding with a hot body would carry that extra energy away at molecular speed until it collided with a colder body which might be a very long way away: heat would be conducted with molecular speed over large distances. We know this does not happen. Thermal energy is transported in a *diffusive* manner, hot molecules colliding with cooler ones and giving them extra energy which they in turn hand on in later collisions. Again, if there were no intermolecular collisions, a gas could not carry sound waves because the molecules would carry the wavemotion off in all directions and rapidly smear it out. Thus, there are properties in which intermolecular collisions play an essential role. We shall first calculate how far molecules travel between collisions and then see how collisions affect thermal conductivity and viscosity.

The average distance that a molecule travels before colliding with another is known as the *mean free path*. What determines whether two molecules will collide? Thinking of the molecules as spheres of diameter S, we see that two molecules will collide if the line along which the centre of one moves passes within S of the

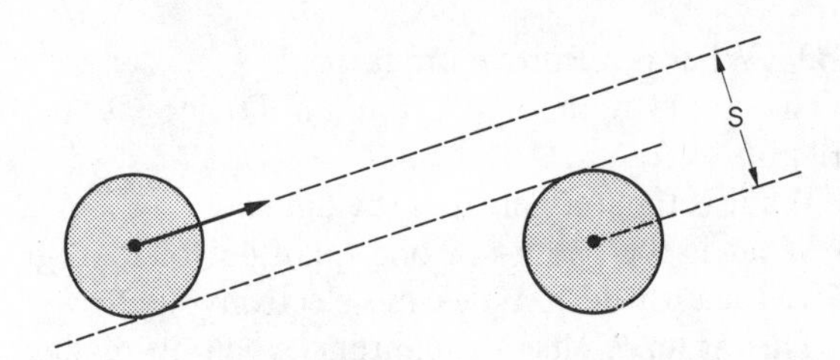

Figure 4.1 Two molecules will collide if their motions bring their centres within one molecular diameter of one another

centre of another (figure 4.1). The effective area swept out by one molecule is therefore πS^2. Then,

$$\text{volume swept out per second} = \pi S^2 \langle c \rangle$$

$$\text{number of collisions per second} = \pi S^2 \langle c \rangle n$$

$$\text{average distance to a collision} = \langle c \rangle / \pi S^2 \langle c \rangle n$$

so

$$l = 1/\pi S^2 n \qquad (4.1)$$

where l is the mean free path and n the molecular density.

For air at s.t.p.,

$$n = N_A/V_m \approx 2{\cdot}7 \times 10^{25}\ \text{m}^{-3}$$

This gives a mean intermolecular distance*

$$R \approx n^{-1/3} \approx 3.3 \text{ nm}$$

The effective molecular diameter, which can be deduced from measurements of thermal conductivity or viscosity (sections 4.2 and 4.3) or estimated from the molar volume of the liquid, is 0.36 nm. This gives $l \approx$ 90 nm. Thus, for air at s.t.p. we may compare the lengths

$$S \approx 0.36 \text{ nm}$$

$$R \approx 3.3 \text{ nm}$$

$$l \approx 90 \text{ nm}$$

The molecules are about ten diameters apart and they travel about 300 diameters before colliding.

l is proportional to temperature at constant pressure and inversely proportional to pressure at constant temperature. A useful result to remember for a quick estimation of mean free paths is that for air the mean free path is about $\frac{1}{4}$ mm when the pressure is $\frac{1}{4}$ mmHg.

4.2 THERMAL CONDUCTIVITY IN GASES

Processes in which some quantity flows or is carried from place to place are known as *transport phenomena.* The first transport property we shall analyze is thermal conductivity where the quantity being transported is thermal energy. We shall use a qualitative argument to derive the form of the expression for thermal conductivity and quote the numerical factor of proportionality. To derive the numerical factor it is necessary to examine in detail the random motion of the molecules. An account of this more rigorous treatment is given in the Appendix.

Consider a gas in which temperature increases in the z direction at a rate $\mathrm{d}T/\mathrm{d}z$. This will result in a flow of heat in the $-z$ direction (from hotter to cooler). We deduce the form of the thermal conductivity by working out the net thermal energy crossing the plane $z = 0$ (figure 4.2).

* If the molecules were arranged in a cubic array (equally spaced a distance R apart in rows parallel to the x, y and z directions, see figure 5.9, page 80), the volume per molecule would be R^3, and the number per unit volume would then be exactly $1/R^3$. They will not, of course, be spaced regularly in this way, but the mean molecular density must still be of this order.

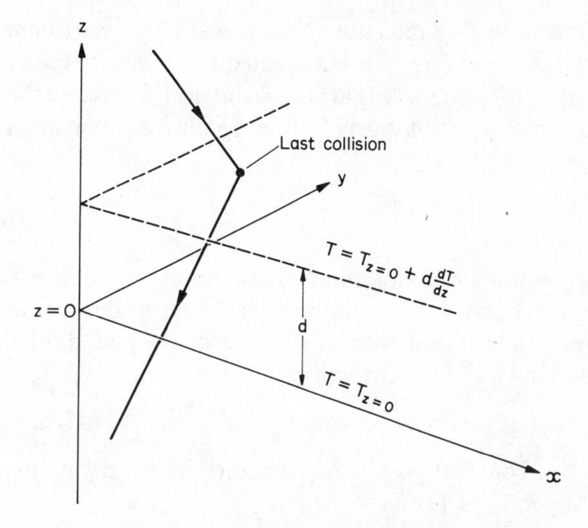

Figure 4.2 Calculation of thermal conductivity of a gas

The thermal energy is carried by the molecules. There is a net flow because molecules crossing $z = 0$ in the downward direction will have made their last collision higher up where the temperature is higher and hence, on the average, will have a slightly greater kinetic energy than those crossing $z = 0$ in the upward direction which made their last collision lower down where the gas is cooler. The 'temperature' of those crossing downward will be $T + d(\mathrm{d}T/\mathrm{d}z)$ where d is the mean vertical distance since the last collision. Hence, the mean thermal energy carried by each molecule crossing downwards will be equal to

$$\left(T + d\frac{\mathrm{d}T}{\mathrm{d}z}\right)c_V'$$

where c_V' is the heat capacity *per molecule.* (It is the heat capacity at constant volume because we only want to include thermal energy associated with the motion of the molecule. c_p' would include work in expansion.)

Similarly, the mean energy carried upwards will be equal to

$$\left(T - d\frac{\mathrm{d}T}{\mathrm{d}z}\right)c_V'$$

The mean net energy transported upwards per molecule is proportional to the difference between these:

$$-d\frac{\mathrm{d}T}{\mathrm{d}z}c_V'$$

The minus sign is present because increase of temperature in the z direction ($\mathrm{d}T/\mathrm{d}z$ positive) causes heat to flow in the $-z$ direction. Now, d will be proportional to the mean free path l (not equal to it because molecules will be travelling at various angles). Therefore the mean energy transported per molecule is proportional to

$$-l\frac{\mathrm{d}T}{\mathrm{d}z}c'_V$$

The number of molecules crossing unit area of $z = 0$ per second will be proportional to the density of molecules times their mean speed. (If you double either, twice as many will cross.) Hence,

$$\text{rate of heat flow per unit area} \propto -n\langle c\rangle lc'_V(\mathrm{d}T/\mathrm{d}z)$$

The constant of proportionality derived in the Appendix is 1/3, so

$$\text{rate of heat flow per unit area} = -\tfrac{1}{3}n\langle c\rangle lc'_V\,(\mathrm{d}T/\mathrm{d}z)$$

Thermal conductivity λ is the constant of proportionality between rate of heat flow per unit area and temperature gradient (page 31):

$$\text{rate of heat flow per unit area} = -\lambda(\mathrm{d}T/\mathrm{d}z)$$

Comparing this with the previous expression for heat flux, we find

$$\lambda = \tfrac{1}{3}nl\langle c\rangle c'_V \qquad (4.2)$$

nc'_V is the heat capacity per unit volume of the gas which is also given by $c_V\rho$ where c_V is specific heat capacity and ρ density. Substituting in terms of these variables we obtain an alternative form:

$$\lambda = \tfrac{1}{3}\rho l\langle c\rangle c_V \qquad (4.3)$$

Thermal conductivity of gases is very small. We may estimate λ for air at s.t.p. as follows. Nitrogen and oxygen are diatomic, so $c'_V = \frac{5}{2}k$ (page 50)

$$\langle c\rangle \approx c_{\mathrm{r.m.s.}} \approx 500\ \mathrm{m\,s^{-1}}\ \text{(page 61)}$$

$$l \approx 90\ \mathrm{nm}\qquad \text{(page 60)}$$

$$n \approx 2.7\times 10^{25}\ \mathrm{m^{-3}}\ \text{(page 60)}$$

which give

$$\lambda \approx 1.4\times 10^{-2}\ \mathrm{W\,m^{-1}\,K^{-1}}$$

This compares with (see section 2.6)

$$\text{typical non-metallic solid, } \lambda \approx 1\ \mathrm{W\,m^{-1}\,K^{-1}}$$

$$\text{typical metal, } \lambda \approx 10^2\ \mathrm{W\,m^{-1}\,K^{-1}}$$

How does the thermal conductivity of a gas depend on pressure and temperature? According to *4.1*, $l \propto 1/n$. Substituting in *4.2* we find the conductivity to be independent of the density of molecules! It seems that something must be wrong because if there is no gas there, there certainly cannot be any thermal conduction! How can the conductivity be independent of density? As the density increases, there are more molecules carrying the heat, but they do not carry it so far because the collisions are closer together. That is where the cancellation occurs. But what happens when we go on *reducing* the pressure? These two terms will continue to cancel until the mean free path becomes so long that the molecules go directly from the hot surface to the cold surface without colliding on the way. The free path cannot become longer than that. In *4.2*, l must then be replaced by a constant of the order of the size of the apparatus. n, of course, continues to decrease as the pressure is lowered so that the thermal conductivity *does* now fall with pressure (figure 4.3). This change from independence of pressure to decreasing with pressure normally occurs at quite low pressures. If the size of the apparatus is a few centimetres, the pressure must be of the order of 10^{-3} mmHg.

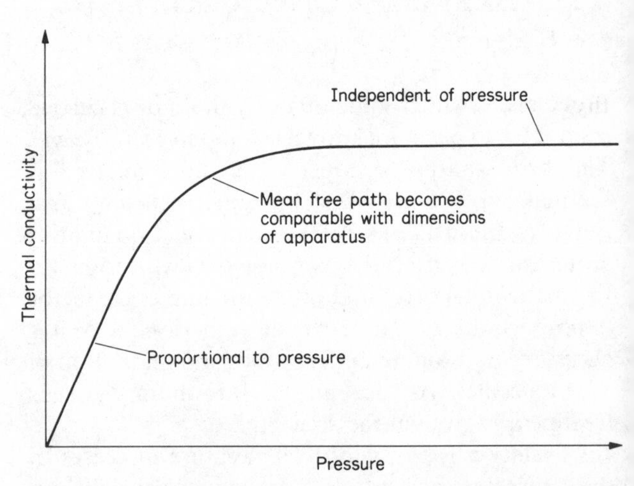

Figure 4.3 Pressure dependence of thermal conductivity of a gas. Viscosity (section 4.3) behaves in the same way.

While thermal conductivity is independent of pressure, the only other term in *4.2* which can cause the thermal conductivity to change is $\langle c\rangle$ which goes as $T^{1/2}$. Thus, provided the pressure is not too low, thermal conductivity of a given gas depends only on temperature as

$$\lambda \propto T^{1/2}$$

Since the thermal conductivity of gases is so small, techniques appropriate to bad conductors have to be used to measure it. In one respect, however, gases (and liquids) are different from solids: because they are fluid, convection can occur. To avoid convection, the apparatus must be arranged so that temperature increases with height in the gas; this gives a gravitationally stable situation with the more dense fluid below the less dense. Figure 4.4 shows a suitable

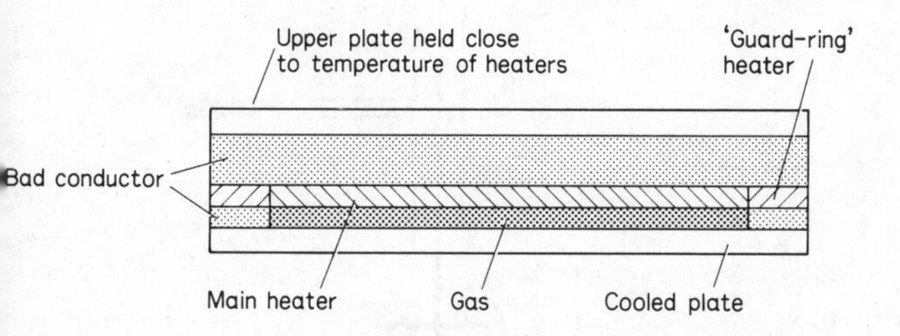

Figure 4.4 An arrangement for measuring the thermal conductivity of a gas

arrangement which is similar to the apparatus shown in figure 2.25 for a bad solid conductor. Here, however, the gas is only present in the space below the heaters so that the temperature increases with height and there is no convection. Above the heaters is a thick layer of a bad conductor (like expanded polystyrene) and the upper plate is held close to the temperature of the heaters so that conduction across the space above the heaters is negligible. To help prevent circulation in the lower chamber the discs should be accurately horizontal.

4.3 VISCOSITY IN GASES

The second transport property we shall consider briefly is viscosity. In order to maintain a velocity gradient in a fluid, a steady shearing force has to be applied (figure 4.5). The viscosity η* is the constant of proportionality between the shearing force per unit area and the velocity gradient. If the velocity of flow is in the x direction, and the velocity increases as z increases at a rate $\mathrm{d}v_x/\mathrm{d}z$ (the velocity gradient), then

$$\text{shear force per area} = p_x = \eta\frac{\mathrm{d}v_x}{\mathrm{d}z} \qquad (4.4)$$

* The Greek letter *eta*.

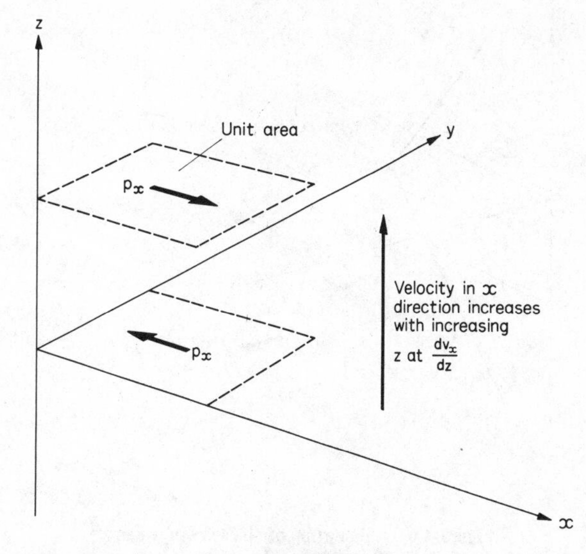

Figure 4.5 To maintain a velocity gradient in a viscous fluid a steady shearing force has to be applied

The shear force is needed because there is exchange of momentum between the faster moving fluid above and the slower fluid below, which tends to speed the slower fluid up and slow the faster down. Thus, the velocity gradient would die away in the absence of the shearing force. To find the viscosity of a gas, we calculate the net momentum transported across the plane $z = 0$ by the molecules crossing in either direction. The argument goes just as for thermal conductivity, but this time the quantity being transported is (horizontal) momentum rather than thermal energy.

Molecules crossing $z = 0$ downwards will, on average, carry extra momentum in the x direction since their last collision was in gas moving faster in the x direction than does the gas at $z = 0$ (figure 4.6). The extra momentum carried will be proportional to $l(\mathrm{d}v_x/\mathrm{d}z)m$ where m is the mass of a molecule. Similarly, molecules crossing $z = 0$ from below will have a horizontal momentum smaller by an amount proportional to $l(\mathrm{d}v_x/\mathrm{d}z)m$. Therefore,

net average momentum per molecule crossing

$$\propto l(\mathrm{d}v_x/\mathrm{d}z)m$$

number crossing unit area per second

$$\propto n\langle c\rangle$$

so,

$$p_x \propto n\langle c\rangle lm(\mathrm{d}v_x/\mathrm{d}z)$$

and, comparing with *4.4*,

$$\eta \propto n\langle c\rangle lm$$

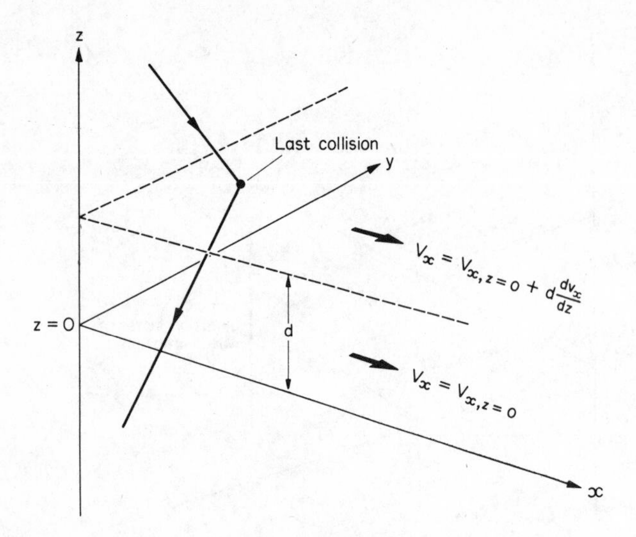

Figure 4.6 Calculation of viscosity of a gas

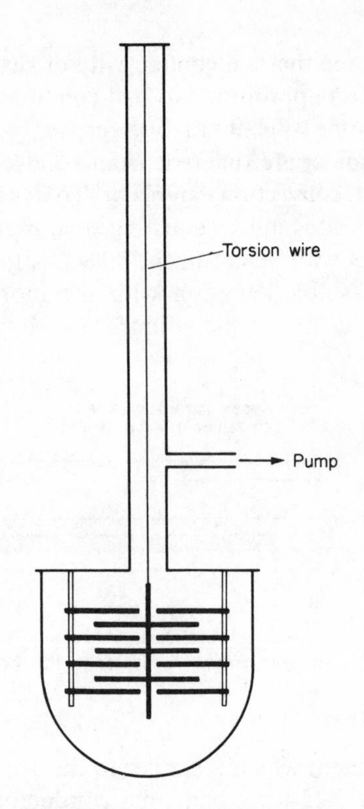

Figure 4.7 Maxwell's apparatus for measuring the viscosity of gases

The constant of proportionality is the same as for thermal conductivity (see Appendix), so

$$\eta = \tfrac{1}{3}nm\langle c\rangle l = \tfrac{1}{3}\rho\langle c\rangle l \qquad (4.5)$$

Comparing with *4.2* we find that thermal conductivity and viscosity are simply related:

$$\lambda = \eta c_V'/m = \eta c_V \qquad (4.6)$$

where c_V is the specific heat capacity of the gas.

As with thermal conductivity, the density cancels between n and l and viscosity is independent of pressure until the pressure becomes so low that the mean free path becomes comparable with the size of the apparatus. This result of kinetic theory was verified in experiments by Maxwell using the apparatus sketched in figure 4.7. A set of discs, hanging from a torsion wire, interleaves with stationary discs. The damping of torsional oscillations is proportional to the viscosity of the gas. As the gas was pumped out of the apparatus, the damping was unchanged until the pressure was low enough for the mean free path to become comparable with the distance between the plates. At higher pressures, the only term in *4.5* causing viscosity to vary is $\langle c\rangle$, and $\eta \propto T^{\frac{1}{2}}$.

4.4 CHANGE OF STATE

There are three general classes of matter: solid, characterized by rigidity and high density; liquid, characterized by similar density, but fluid; gas, again fluid but much less dense. Most substances can exist in all three forms, and which form will be present at any time will depend on the conditions of pressure and temperature. In the next section we shall describe typical behaviour; but it will help us to understand why substances behave the way they do if we first think about the energetics of change of state.

The existence of solid, liquid and gas as distinct states of matter results from the attractive forces which always act between atoms and molecules. Because they attract one another, when two are brought close together, their potential energy decreases (as does the potential energy of a mass when it is lowered in the gravitational attraction of the earth). The state of lowest potential energy is when they are close together. (See section 5.1.) Whether or not they remain close together depends on temperature. Change of state is determined by competition between thermal energy, which acts to excite random thermal motion, and attractive energies, which act to draw molecules together into a condensed state.

Consider a gas at a relatively high temperature. Because the temperature is high the molecules will have large kinetic energies. When two come close together, there will be the reduction in potential energy which acts to try to hold them together; but if thermal energies are much greater than this, the attraction is not strong enough to keep them together. It is like rolling a ball into a hollow in the ground: if it is rolled in fast enough, it simply continues out the other side. As the temperature is lowered, however, the translational kinetic energies fall and eventually molecules tend to stay bound together by the attractive forces. When this happens, liquid condenses out. Within the liquid, the molecules are still mobile. Only at still lower temperatures do they finally settle into a rigid ordered arrangement corresponding to the solid. In both condensed states the molecules are close together. A solid melts when the thermal motions are sufficient to enable the molecules to slide past one another. The extra space they require to do this is small and that is why there is only a small increase in volume on melting.

Condensation to liquid (and, to a lesser extent, solidification) are affected by pressure. As the pressure is raised, the substance is discouraged from being in the state which takes up more volume. Thus, increase of pressure encourages condensation from gas to liquid. It also promotes solidification in substances which contract on freezing, which is what usually happens. Water is an exception in that it expands on freezing (because the molecules arrange themselves into a special structure in ice in which they take up more space). Thus, with ice, increased pressure promotes melting. Melting of ice by application of pressure is known as *regelation* and is partly responsible for low friction in skiiing and skating.

4.5 *p-V-T* RELATIONSHIPS FOR PURE SUBSTANCES

Figure 4.8 shows a typical series of isotherms for a pure substance. At high temperatures, like T_1, energies associated with intermolecular attraction are negligible compared with thermal energies and the substance is gaseous. If the temperature is high enough, the relationship between p and V is similar to the rectangular hyperbola, $pV =$ constant, characteristic of an ideal gas. Only at very small volumes (corresponding to extremely high pressures) are the molecules close enough for their size to matter. The pressure then increases as the volume is reduced faster than it would if the molecular volume were negligible (figure 4.9). At lower temperatures, like T_2, as volume is decreased, one reaches the condition at A where it becomes energetically favourable for liquid to condense out. Now the saturated vapour pressure of a liquid (section 5.6) depends only on temperature, so, as the volume is reduced further, the pressure stays constant while more liquid condenses until, at B, all the substance is liquid. The pressure then rises very rapidly because, compared with the gas, the liquid is rather incompressible.

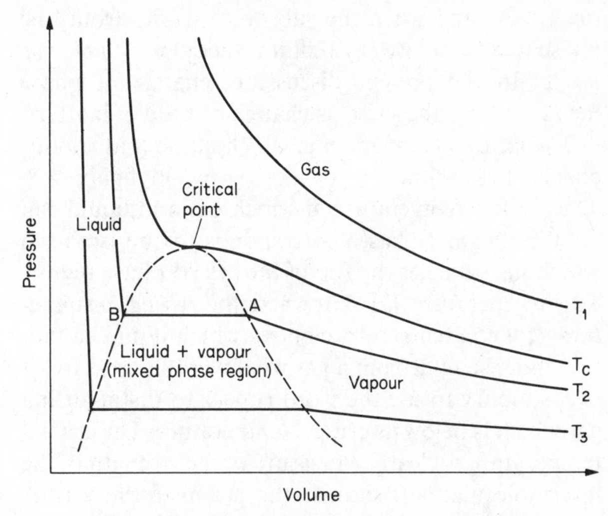

Figure 4.8 Typical isotherms for a pure substance

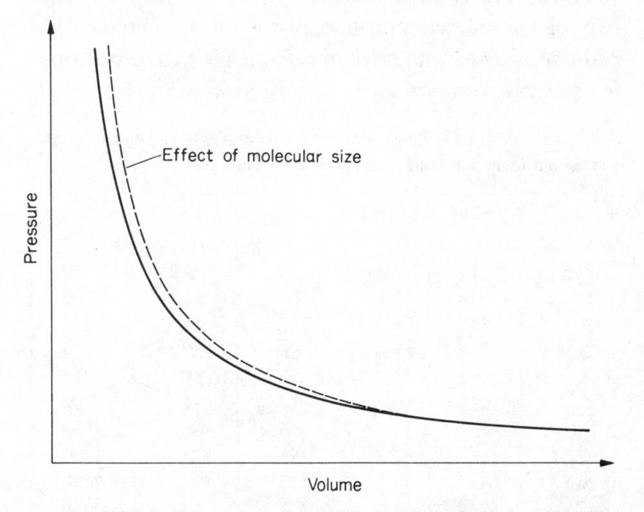

Figure 4.9 Molecular size causes pressure to rise more rapidly as volume is reduced because it reduces the space available for molecules to move in

When a system has distinct parts, as when both liquid and gas are present, we say that it has more than one *phase*, and when the substance passes from one phase to another we say that it undergoes a *change of phase*. In the present discussion, change of phase means exactly the same as change of state of matter.

The *mixed phase region* in which liquid and vapour coexist lies below the broken curve in figure 4.8. The highest temperature at which gas and liquid can exist as distinct phases corresponds to the isotherm which just touches the top of the mixed phase region. This temperature T_c is known as the *critical temperature*. It is the temperature above which liquid cannot be condensed out from a gas whatever the pressure. It is customary to use the word *vapour* to distinguish a gas which is below its critical temperature. The critical temperature is clearly a measure of the strength of the intermolecular attractions: the stronger the attractions, the higher the critical temperature will be. The pressure and volume at which liquid just appears (the top of the mixed phase region) are known as the *critical pressure* and *critical volume*. Critical constants for several common gases are listed in table 4.1.

Table 4.1 Normal boiling points, critical constants and approximate maximum inversion temperatures of common gases

Gas	T_b/K	p_c/atm	V_c/ml mol^{-1}	T_c/K	$T_{i,\max}$/K
^{4}He	4.22	2.26	58	5.26	50
H_2	20.4	12.8	65	33.3	200
Ne	27.1	26	42	44.5	210
N_2	77.4	33	90	126	620
A	87.3	48	75	151	720
O_2	90.2	50	74	154	890
CO_2	195^s	73	96	304	1500
C_3H_8	231	42	195	370	1800
SO_2	263	78	123	430	2000

^{s} This is the sublimation temperature. At 1 atm, the solid passes directly into vapour.

Since p and V depend on each other in a unique way for each temperature, we could represent the behaviour of the substance in a three-dimensional plot in which the three axes are p, V and T. The relationship between the three variables is then represented by a surface. This is sketched in figure 4.10. In this representation, isotherms are lines on the surface for which T is constant. Those labelled T_1, T_c, T_2 and T_3 correspond to the isotherms of figure 4.8.

We see in figure 4.10 that, besides the liquid + vapour mixed phase region, there are also mixed phase regions in which solid + liquid and solid + vapour are present. In the former, compression of the liquid at constant temperature causes solidification. (The diagram is different for water where compression promotes melting.) In the latter, which is confined to lower temperatures like T_4, isothermal compression of the vapour causes it to condense directly into the solid. The reverse process, change of phase directly from solid to vapour, is called *sublimation*. Iodine crystals sublime readily if warmed at atmospheric pressure.

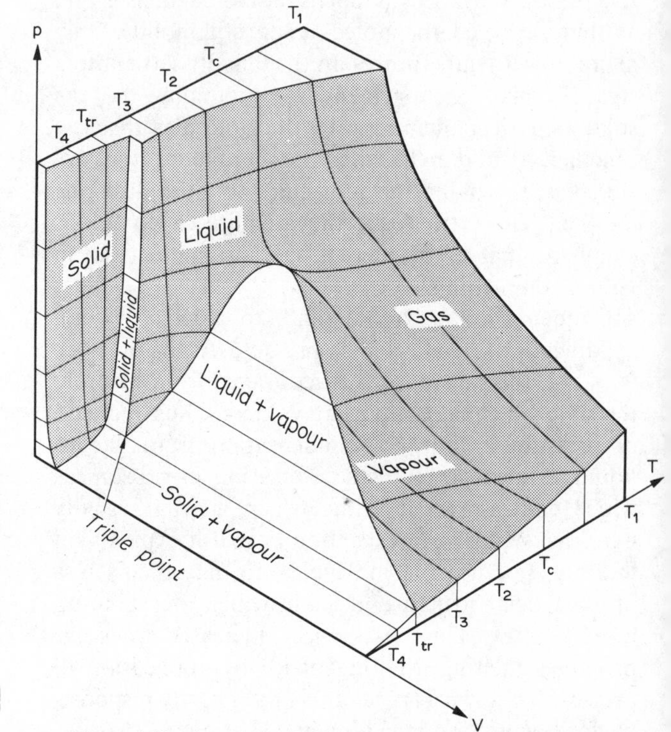

Figure 4.10 A typical *p-V-T* relationship for a pure substance

We have already pointed out that the top of the liquid + vapour mixed phase region defines unique values of pressure, volume and temperature for each substance, the critical constants. We see from figure 4.10 that there is another point which is uniquely defined for each substance: the point where the three mixed phase regions meet. Here, solid, liquid and vapour are simultaneously in equilibrium with one another. This is the *triple point*. It is the triple point of water which is used to define the kelvin.

4.6 NON-IDEAL GASES

Clearly, a complicated relationship like that shown in figure 4.10 cannot be represented by any simple formula. It is, however, possible to develop relatively simple formulae to give a rough representation of how real substances behave as the temperature is lowered towards the critical temperature (i.e. in the gas region where the behaviour is becoming non-ideal). The best known of these is the *van der Waals' equation*:

$$(p + a/V_m^2)(V_m - b) = RT \qquad (4.7)$$

a and b are constants which take account of the effects of intermolecular attraction and molecular volume respectively. We may explain the equation as follows.

$(V_m - b)$: Because the molecules have finite volume, the space available for any one molecule to move in is reduced. The reduction is proportional to the number of molecules present. Hence, for a given mass of gas, a constant has to be subtracted from the total volume. The units of b are $m^3\,mol^{-1}$ (the same as those of V_m from which it is subtracted).

$(p + a/V_m^2)$: A molecule about to hit the wall will be slowed up by the attraction of the molecules behind it in the mass of the gas (figure 3.6), and so, on impact, will exchange less momentum with the wall. The number of molecules pulling it back will be proportional to the density, i.e. inversely proportional to the molar volume. Thus, each molecule will have its momentum reduced as it approaches the wall proportionally to V_m^{-1}. But the number of molecules hitting the wall is also proportional to the density, and so to V_m^{-1}. Hence, the total reduction in momentum and therefore in pressure is proportional to V_m^{-2}. The constant of proportionality a has units $J\,m^{-3}\,mol^{-2}$ (so that a/V_m^2 has the same units as pressure to which it is added).

Figure 4.11 shows van der Waals' isotherms. The general shape is qualitatively correct down to the critical temperature. Below the critical temperature the equation gives curves which loop and have a region of positive slope. This region has no physical meaning because a substance whose pressure *increases* as the volume increases would be mechanically unstable: the bigger it got, the harder it would push! Instead, one must assume that the looped region should be crossed by a horizontal line (broken) representing the condensation from vapour to liquid. (Thermodynamics shows that the line should be drawn so that the areas of the loops above and below it are equal.)

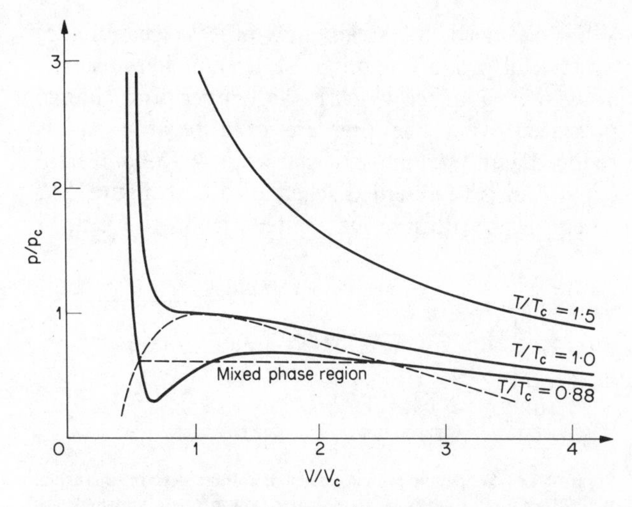

Figure 4.11 Van der Waals' isotherms. The curves are plotted in dimensionless form, the pressure, volume and temperature being measured in units of the critical pressure p_c, the critical volume V_c and the critical temperature T_c.

Van der Waals' equation is quite useful at higher temperatures where deviations from ideal gas behaviour are not too great; but at lower temperatures it only gives a very rough indication of how real substances behave, which is not surprising in view of the relatively simple arguments on which it is based. (The arguments can be developed more rigorously than we have done here!)

4.7 LIQUEFACTION OF GASES

Intermolecular attraction, represented by the term a/V_m^2 in van der Waals' equation, can be thought of as an *internal pressure* against which work has to be done when the gas expands: energy has to be supplied because the potential energy of the molecules must increase as they move further apart. If the intermolecular attractions are significant, a real gas will not obey Joule's law (page 40). In a free expansion the total internal energy is constant. But the total internal energy is made up of two parts: intermolecular potential energy, and energy of thermal motion. If the former has to increase, the latter must decrease to keep the total constant; i.e. the temperature will fall. Real gases (except in the low density limit when their behaviour approaches that of the ideal gas) will therefore *always* cool in a Joule expansion.

However, a Joule expansion is not much good for liquefying a gas because it is a once-for-all process.

What one needs is a steady flow process (section 2.5) which will produce cooling. Many liquefiers use the *Joule–Kelvin effect* which is the temperature change produced when the pressure of a flowing gas is reduced without doing external work. A Joule–Kelvin expansion is represented schematically in figure 4.12 where high pressure gas enters from the left, it

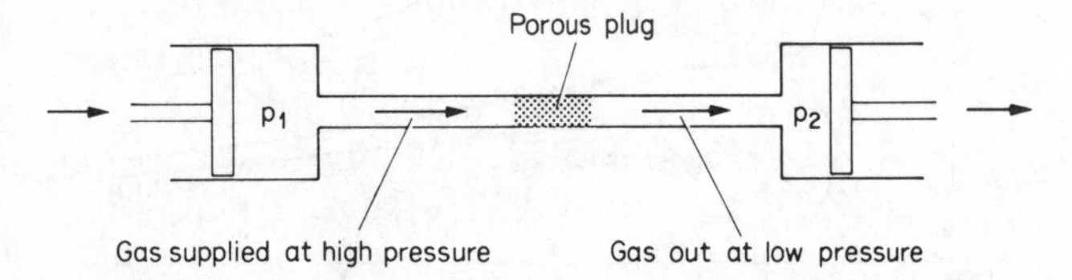

Figure 4.12 Schematic representation of a Joule–Kelvin expansion. A Joule–Kelvin expansion is a steady flow process in which the pressure of the gas is reduced without doing external work.

passes through a porous plug (or small orifice) which limits the flow and drops the pressure, and leaves at low pressure on the right. This is a steady flow process in which no external work is done, no heat enters or leaves the system and kinetic and potential energies are negligible. The general equation for flow processes (equation *2.21*, page 29) simplifies to

$$u + pv = \text{constant} \qquad (4.8)$$

Now u = constant is just the condition of the Joule expansion we have discussed above, so that the u term always contributes a cooling effect. Whether the gas cools or warms in a Joule–Kelvin expansion depends on the behaviour of pv, and this can go either way. If the temperature is high so that intermolecular attractions have little effect, then only molecular volume matters. This reduces the volume available for the molecules to move in so that pv increases as the gas is compressed at constant temperature (figure 4.9). On the other hand, if the temperature is low enough for the intermolecular attractions to be more important, then attractions will increase as the molecules get closer and pv will decrease as the gas is compressed. Typical behaviour of pv at constant temperature is shown in figure 4.13. Thus, if the temperature is high, the pv term in equation *4.8* will contribute a warming as the gas expands, and, if this contribution is large enough, the *net* effect will be a warming. At lower temperatures, the effects leading to cooling will predominate and the gas will fall in temperature. Whether cooling or warming occurs

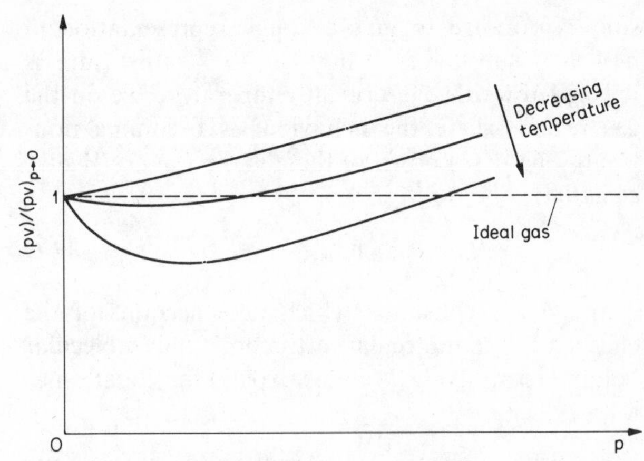

Figure 4.13 Typical dependence of the product *pv* on pressure

also depends on pressure. There is therefore, for each gas, a *maximum inversion temperature* above which a Joule–Kelvin expansion *always* gives a warming and below which it *can* give a cooling. Maximum inversion temperatures and normal boiling points are included in table 4.1 (page 66).

A single Joule–Kelvin expansion may not produce enough cooling for liquid to form; but the efficiency

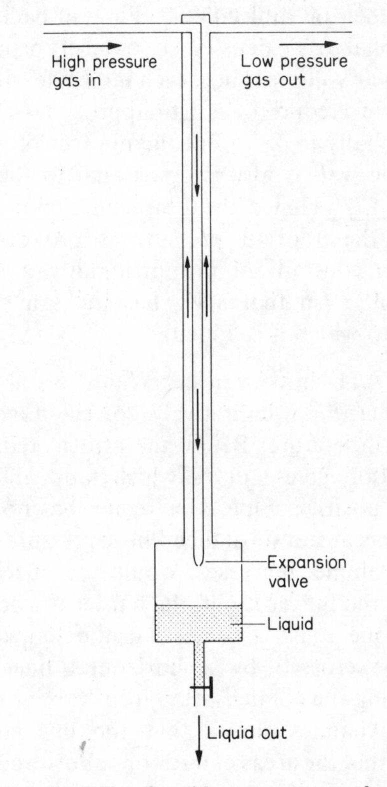

Figure 4.14 A simple liquefier with a countercurrent heat exchanger

can be increased if the high pressure incoming gas is cooled by the low pressure exhaust gases before making the expansion. This is done in a *countercurrent heat exchanger*: the gases flowing in opposite directions are in thermal contact so that heat is exchanged. A simple liquefier with a heat exchanger is illustrated in figure 4.14.

Gases with low maximum inversion temperatures have to be precooled before a Joule–Kelvin expansion will produce further cooling. This is usually done by using other liquid gases, boiling at low temperature, to cool the incoming gas. This is the principle of the *cascade process*. Figure 4.15 shows a photograph of the inside of a helium liquefier using the cascade process. Figure 4.16 shows schematically the stages used in it. Three gases are involved, nitrogen, hydrogen and helium. The first stage of cooling uses liquid nitrogen (which may be bought cheaply) boiling under reduced pressure at about 67 K. This is below the maximum inversion temperature of hydrogen, so, in the next stage, liquid hydrogen is formed by a Joule–Kelvin expansion. Here, the liquid hydrogen boils a little above atmospheric pressure at about 22 K. Some of the liquid is drawn off into another container where it is boiled under reduced pressure at about 14 K, which is well below the maximum inversion temperature of helium. In the final stage the helium is expanded. The pressure falls to atmospheric, and as a result of the cooling, some liquid condenses out at 4.2 K, the normal boiling point of helium. The

Figure 4.15 The interior of the Ashmead helium liquefier
The liquefier was used in the Cavendish Laboratory of Cambridge University to produce liquid helium for research. During operation, an airtight outer casing would be in place and the space inside would be kept at high vacuum for thermal isolation.

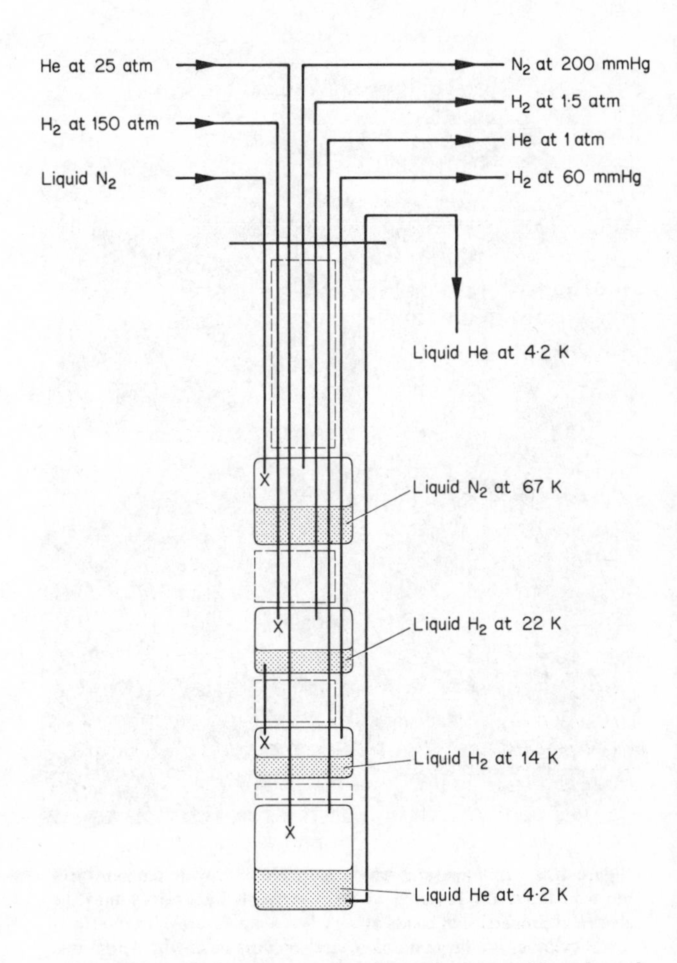

Figure 4.16 A schematic representation of the stages of the Ashmead liquefier, with heat exchangers shown in dotted outline

remaining cold gas returns up the liquefier. Between each stage are heat exchangers so that the up-flowing cold gases precool the down-flowing high pressure gases before expansion.

Gases may also be cooled by making them do external work, as, for example, when expanding against a piston. Such an expansion *always* produces a cooling because the internal energy must fall as the work is done; but this process is not often used at very low temperature, mainly because of the difficulty of designing the moving mechanical parts so that they will operate satisfactorily at low temperatures. However, this method of cooling is used in some helium liquefiers in combination with a Joule–Kelvin expansion to produce liquid directly from high pressure gas at room temperature.

The importance of liquid helium is that helium has the lowest boiling point of any substance. The common isotope ^{4}He boils at atmospheric pressure at 4.2 K; the lighter isotope ^{3}He at 3.2 K. By making them boil under reduced pressure the temperatures can be reduced to about 0.9 K and 0.3 K respectively. Liquid helium therefore provides a convenient means of reaching temperatures below a few kelvins. Reduction of thermal motions by cooling to low temperatures is of importance in many branches of pure and applied physics. Figure 4.17 shows a photograph of an apparatus which uses ^{3}He to cool specimens of special semiconductors for study at very low temperature.

Figure 4.17 An apparatus which uses ^{3}He to provide temperatures down to 0.3 K for research at the Cavendish Laboratory into the electrical properties of solids at very low temperatures. The vertical metal cylinder is a large stainless steel vacuum flask which contains liquid nitrogen boiling at 77 K. Immersed in the liquid nitrogen is another vessel, suspended from above, whose interior is kept at high vacuum for thermal isolation of a smaller vessel within it which contains the liquid helium. The material under study is in thermal contact with the boiling helium. To the right are the pumps for removing the helium gas.

PROBLEMS

Mean free path in a gas; diffusive motion, molecular size

4.1 In the kinetic theory derivation of the ideal gas law it is assumed that the molecules have negligible size and that there are no attractive forces between them. What properties of real substances can only be understood if molecules do have finite size and if there are attractive forces between them? Explain how finite molecular size and intermolecular attraction are involved in each case.

4.2 At s.t.p. the density of helium is 0.178 kg m^{-3} and the mean free path 285 nm.
a) What is the number of molecules per cubic metre?
b) What is the effective atomic diameter?
[For helium, $A_r = 4$.]

4.3 The density of liquid argon is 1.4 Mg m^{-3}.
a) How many argon atoms are there per unit volume in the liquid?
b) What is the approximate atomic diameter? (You may assume cubic packing; see footnote on page 61.)
c) Estimate the mean free path in gaseous argon at s.t.p.
[Relative atomic mass of argon = 40.]

Thermal conductivity of gases

4.4 The thermal conductivity of argon at 20 °C is measured at various pressures (in an apparatus like that shown in figure 4.4) by finding the heat conducted across a layer of the gas 1.0 mm thick. The conductivity

is found to be constant until the pressure is reduced to about 0.08 mmHg.
a) Why does the conductivity remain constant down to this pressure?
b) Why does the conductivity decrease below this pressure?
c) How would the conductivity vary with pressure at low pressures?
d) What is the number of molecules per cubic metre at 0.08 mmHg and 20 °C?
e) Estimate the effective atomic diameter of argon. [Argon has relative atomic mass of 40 and density at 20 °C and 1 atm pressure 1·66 kg m^{-3}.]

4.5 Use the following data for ammonia at 20 °C to estimate
a) the mean molecular speed (take $\langle c \rangle \approx c_{\text{r.m.s.}}$).
b) the mean free path,
c) the number of molecules per cubic metre,
d) the effective molecular diameter.

$$M_r = 17$$
$$\lambda = 2.3 \times 10^{-2}\ \text{W m}^{-1}\ \text{K}^{-1}$$
$$\rho = 0.72\ \text{kg m}^{-3}$$
$$c_V = 1.63\ \text{kJ kg}^{-1}$$

Viscosity of gases

4.6 What is meant by the *viscosity* of a fluid? Explain in terms of the movements of molecules why a body moving in a gas experiences a viscous drag. Show that the SI unit of viscosity is Pa s.

4.7 Between a flat metal plate and a smooth plane is a film of oil 0.3 mm thick. If the viscosity of the oil is 0.5 Pa s, and the plate has a mass of 100 g and an area of 100 cm^2, how fast does the plate move down the plane when the plane is inclined at 10° to the horizontal?

4.8 Use the following data to estimate the viscosity of nitrogen at s.t.p.

Density of gas at s.t.p. = 1.25 kg m^{-3}
Molar volume of liquid = 3.54 × 10^{-5} m^3
Relative molecular mass = 28

4.9 Given that the relative atomic mass of helium is 4.00, what can you deduce from the observations that at s.t.p.

$$\lambda_{\text{He}}/\lambda_{\text{Ne}} = 3.12 \quad \text{and} \quad \eta_{\text{He}}/\eta_{\text{Ne}} = 0.626?$$

Microscopic aspects of change of state

4.10 Discuss the energy changes associated with melting and evaporation. Why does pressure always promote condensation and generally promote solidification?

p-V-T relationships for pure substances: change of phase, critical constants, triple point

4.11 Give an account of how pressure, volume and temperature are related for a typical pure substance.
Why does one generally expect that a substance with a low critical temperature will also have a low boiling point?

Non-ideal gases; approximate equations of state

4.12 An imperfect gas, whose behaviour is well described by van der Waals' equation, is used in a constant volume gas thermometer. Will it give true thermodynamic temperatures if calibrated
a) at the triple point, putting

$$T = T_{tr}\frac{p}{p_{tr}}$$

b) at ice and steam points, putting

$$T = 273.15 + \frac{99.97(p - p_i)}{(p_s - p_i)}$$

4.13 A gas obeys van der Waals' equation with values for the constants, $a = 2.48 \times 10^{-2}$ N m^4 mol^{-2} and $b = 2.66 \times 10^{-5}$ m^3 mol^{-1}. 0.1 mol of the gas occupies 5×10^{-5} m^3 when the pressure is 50 atm. (a) What is the temperature? The gas is now compressed at constant temperature until its volume is 2×10^{-5} m^3. (b) What is the pressure?

Liquefaction of gases: cooling in free and Joule–Kelvin expansions; maximum inversion temperature; heat exchangers; cascade liquefaction

4.14 Discuss the principles involved in the liquefaction of gases.

4.15 What is a *Joule–Kelvin expansion*? How does the temperature of an ideal gas change in a Joule–Kelvin expansion?

4.16 Why do real gases

a) always cool when making an adiabatic reversible expansion (i.e. when expanding against a piston)?

b) always cool when making a free (Joule) expansion?

c) sometimes cool when making a Joule–Kelvin expansion?

5 Interatomic Forces

In this chapter we shall look at interatomic forces in a little more detail and see how their nature can explain many simple properties of materials.

5.1 INTERATOMIC POTENTIAL ENERGY

We may easily work out how the force between two atoms must depend on the distance between them. At very large distances the force must vanish because all types of force decrease with distance. As the atoms come closer, the attractive force will increase; but when the atoms are very close, when they are pressed hard together, the force must be repulsive. At some intermediate separation, typically a few tenths of a nanometre, the attraction and repulsion balance, and this is the equilibrium separation r_0. The form of the relationship between force F and distance r is sketched in figure 5.1. The sign convention is that a positive force is in the direction of increasing r and is therefore a repulsion. A negative force is in the direction of decreasing r and is therefore an attraction.

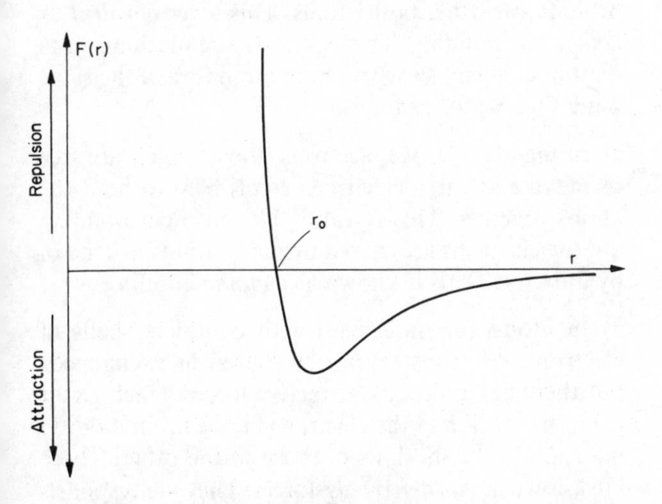

Figure 5.1 How interatomic force depends on atom separation

It is usually convenient to discuss interatomic forces in terms of the corresponding potential energy $\phi(r)$ which is the work done in bringing up the second particle from infinity to a distance r from the first. Mathematically, ϕ and the force F are related by

$$F(r) = -\frac{\mathrm{d}\phi}{\mathrm{d}r}$$

or $\qquad$ (5.1)

$$\phi(r) = -\int_{\infty}^{r} F(r)\,\mathrm{d}r$$

The potential energy corresponding to the force sketched in figure 5.1 is shown in figure 5.2. Here, the

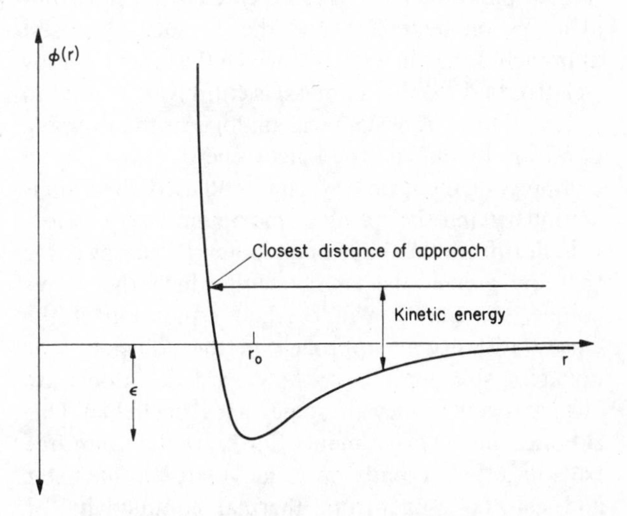

Figure 5.2 Interatomic potential energy as a function of atom separation corresponding to the force shown in figure 5.1

equilibrium separation corresponds to the minimum in the curve: work has to be done on the system to reduce the interatomic distance below r_0 because, as the separation is reduced, the force is repulsive; similarly, work has to be done on the system to increase the separation to greater than r_0 because, in this region the force is attractive. The advantage of referring to the potential energy curve is that it is often easier to visualize how the system behaves by imagining a mechanical analogue: if a track in the shape of the potential energy curve is set up vertically and a ball runs in the track, gravitational potential replaces interatomic potential and the motion of the ball when displaced from the position of minimum energy is equivalent to the relative motion of the atoms when they are displaced. For example, if moved slightly from the minimum and released, the ball would oscillate about the equilibrium position. This corresponds to oscillation of the interatomic distance r in a diatomic molecule.

We should note that the effective 'size' of the atoms will depend on what experiment we do to measure the size. The minimum in the potential energy curve corresponds to the equilibrium separation of the atoms when bound together and gives us one measure of atomic size: if the atoms were identical we would say that r_0 was equal to two atomic radii since r_0 is measured from atom centre to atom centre. But if two atoms collide in a gas, the apparent size is smaller: they approach each other with a positive total energy (kinetic plus potential), as represented by the horizontal arrow on figure 5.2, and the distance of closest approach is the separation at which the kinetic energy (relative to the centre of mass) is entirely converted to potential form. A mechanical analogue is the stopping of a train by buffers: the kinetic energy is used up in compressing the spring; when it is all used, the train is stationary and the spring at maximum compression. It is clear from the shape of the potential energy curve that the greater the energy with which the atoms collide, the smaller will be their separation at the closest distance of approach in the collision. The apparent size depends on how hard the atoms are pushed together (they are somewhat squashable). This is borne out by experiments: for example, mean free path in a gas depends on effective atomic diameter and can be found from thermal conductivity or viscosity (sections 4.2 and 4.3); careful measurements show that the molecular radius appears to decrease as the temperature rises and the collisions become more violent.

What are the origins of the forces? The *repulsive forces* usually result from overlapping of the electrons belonging to the two atoms. According to the *Pauli exclusion principle*, two electrons cannot occupy the same state. (This accounts for the shell model of the atom.) As a result, when the atoms are brought together so that their electrons would begin to overlap and share the same region of space, the electron clouds have to be distorted (squashed) so that the exclusion principle is not violated. This costs energy, so work has to be done in bringing them together which means that there is a repulsive force.

There are several types of attractive force:

a) If the atoms are oppositely charged there is simple *electrostatic* attraction obeying the inverse square law. This produces what is known as *ionic* bonding. It is the attractive force in the alkali halides: NaCl, for example. In the solid, the positive and negative ions are arranged alternately in a three-dimensional cubic array (figure 5.3).

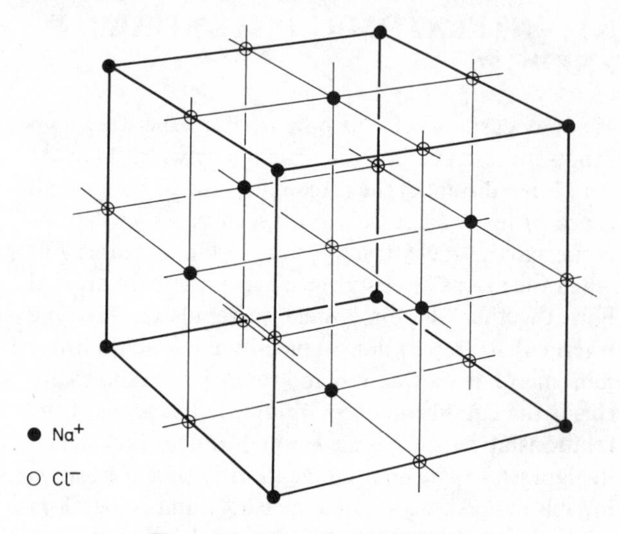

Figure 5.3 The sodium chloride lattice

b) Atoms can share electrons; the electrons go into orbitals enclosing both atoms. This gives *covalent* or *homopolar* bonding. This is always the mechanism in diatomic molecules where both atoms are of the same kind, O_2, N_2 for example.

c) In metals, the free electrons (those which are free to move and carry electric current) help to hold the atoms together. This is a little like covalent bonding but the electrons are shared by many atoms instead of by only two. This is known as *metallic* bonding.

d) In atoms (or molecules) with complete shells of electrons, electrons cannot be shared or exchanged; but there are still weak attractive forces which result from the motion of the electrons in one atom modifying slightly the motions of those in the other. These are known as *van der Waals'* forces. They are responsible for condensation of the inert gases.

Table 5.1 Typical bond energies

Type	*Ionic*	*Covalent*	*Metallic*	*Van der Waals*
ε/J	5×10^{-19}	5×10^{-19}	3×10^{-19}	2×10^{-21}
ε/eV	3	3	2	0.01
εk^{-1}/K	3×10^4	3×10^4	2×10^4	100

The strength of a bonding force is represented by the depth of the resulting potential hollow, labelled ε in figure 5.2. Typical orders of magnitude of ε in the four cases are listed in table 5.1 in joules, electronvolts and kelvins. The electronvolt, the energy of an electron which has been accelerated through a potential difference of one volt, is a useful unit for discussing atomic energies which generally come out of order unity. Since the charge on the electron is $e = -1.6 \times 10^{-19}$ C,

$$1 \text{ eV} = 1.6 \times 10^{-19} \text{ J}$$

An energy may be expressed as an *equivalent* temperature by putting $kT = \varepsilon$. Since kT is the order of magnitude of thermal energy, this gives the temperature at which the energy of thermal motion is roughly equal to the potential energy holding the atoms together. It is therefore the temperature which would break up molecules with that energy of binding. In the table we list εk^{-1} in kelvins.

Some students think that *gravitational attraction* is important between atoms. It is utterly negligible, as we can easily show by calculating the gravitational potential energy for a typical case. We will take argon which has $A_r = 40$ and interatomic distance in the solid, $r_0 = 0.38$ nm. The atomic mass is $m = m_u A_r$. The gravitational interatomic force, by Newton's law, is

$$F = -m^2 G/r^2$$

where G is the gravitational constant. The change in gravitational potential in bringing the atoms together from infinity to a distance r_0 apart is

$$\phi(r_0) = -\int_{\infty}^{r_0} F(r)\,\mathrm{d}r = +m^2 G \int_{\infty}^{r_0} \frac{\mathrm{d}r}{r^2} = -m^2 G/r_0$$

Substituting the values of the variables, we find

$$\phi(r_0) \approx -8 \times 10^{-52}\, J \approx -5 \times 10^{-33} \text{ eV}$$

which is more than 10^{30} times smaller than any of the energies in table 5.1!

We will now see how interatomic forces are related to several basic properties of materials.

5.2 ELASTICITY

When a solid is stretched, the atoms are pulled slightly further apart; that is, r is increased slightly above its equilibrium value r_0. Referring to figure 5.1, we see that *for small displacements* the force required between each pair of atoms is proportional to the displacement from equilibrium. Now the total force on a solid is just the sum of the tensions down all the rows of atoms stretching from one end of the solid to the other. In each row of atoms, tension is proportional to extension, so tension must be proportional to extension for the solid as a whole also: solids obey *Hooke's law* (force proportional to extension). The constant of proportionality between tensional force per unit area P and fractional increase in length $\Delta L/L$ is the *Young modulus* E:

tensional force per area = E × fractional increase in length

or, in symbols,

$$P = E\frac{\Delta L}{L} \tag{5.2}$$

Thus, a rod or wire of cross-sectional area A and original length L_0 is stretched by a tensional force F to a length

$$L = L_0\left(1 + \frac{1}{E}\frac{F}{A}\right) \tag{5.3}$$

A simple method of measuring the Young modulus based on the stretching of a wire is illustrated in figure 5.4.

$\Delta L/L$ is dimensionless, so the dimensions of E and P are the same, namely, force per area or pressure. The normal unit for E is therefore the pascal.

In differential notation, E is given by

$$E = L\left(\frac{\partial P}{\partial L}\right)_T$$

The suffix T reminds us that it is assumed in the definition of E that the temperature is kept constant as the material is stretched.

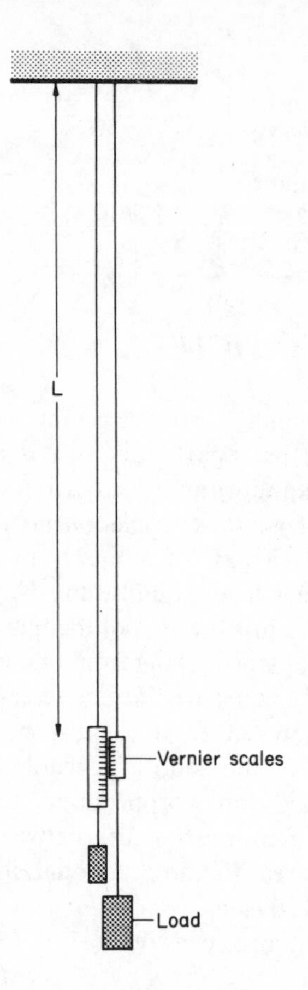

Figure 5.4 Determination of the Young modulus of a wire. Displacements are measured against a scale supported by a similar wire under constant tension. This compensates for any movement of the support and also for expansion resulting from any change of temperature.

Now we have reasoned that tensional force per unit area is directly proportional to fractional increase in length because, for small displacements, the force between any two atoms is directly proportional to the displacement from their equilibrium separation. The same reasoning leads us to conclude that, when hydrostatic pressure is applied to a body, there will also be direct proportionality between increase in pressure and fractional decrease in volume. The only difference here is that hydrostatic pressure acts on a body from all sides whereas, in defining the Young modulus, we considered a force acting along one direction only. The constant of proportionality between increase in pressure and fractional decrease in volume is the bulk modulus K which we have already discussed in section 3.3.

Whenever a body is deformed, the force per unit area producing the deformation is called the *stress*, and the fractional deformation produced is called the *strain*. Thus we always have

modulus = stress/strain

5.3 THERMAL EXPANSION

When two atoms are displaced from their equilibrium separation, the resulting force is directed so as to bring r back to the equilibrium value r_0. Thus, as our mechanical analogue with a ball on a track (page 74) suggested, r can oscillate about r_0. For small displacements, the restoring force is proportional to displacement and oscillations will be simple harmonic. According to the principle of equipartition of energy (page 48), the mean energy of oscillation will be kT ($\frac{1}{2}kT$ in each of potential and kinetic forms). For larger displacements, the oscillations will not be simple harmonic (because the restoring force does not remain proportional to displacement), but we may expect the mean energy still to be of this order. At the extremes of motion, all the energy will be in potential form ($\dot{r} = 0$), so the oscillations will reach to about kT above the minimum in the potential energy curve. Now suppose the energy were symmetric about the minimum (say a parabola), then, as temperature increases, the amplitude of oscillation will increase, but *the mean value of r will be unchanged* (figure 5.5a). But the potential energy of two atoms is not symmetrical about r_0: it slopes up more steeply for $r < r_0$ and less steeply for $r > r_0$. This means that, as the amplitude of oscillation increases, the mean position shifts to larger values of r (figure 5.3b). The atoms take up more room as the amplitude of vibration increases: the substance expands.

Detailed theory shows that, provided the expansion is not too great, the mean separation increases linearly with temperature. In a solid, if each interatomic distance increases linearly with temperature, so will the total length of the solid. The constant of proportionality between the fractional increase in length and the increase in temperature is the *linear expansivity* α:

fractional increase in length = α × increase in temperature

or, in symbols,

$$\frac{\Delta L}{L} = \alpha \, \Delta T \qquad (5.4)$$

Thus, for an increase of temperature t, a solid of original length L_0 expands to

$$L = L_0(1 + \alpha t) \qquad (5.5)$$

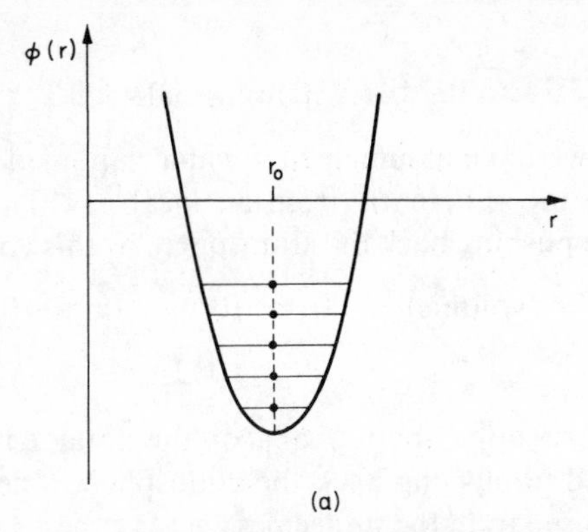

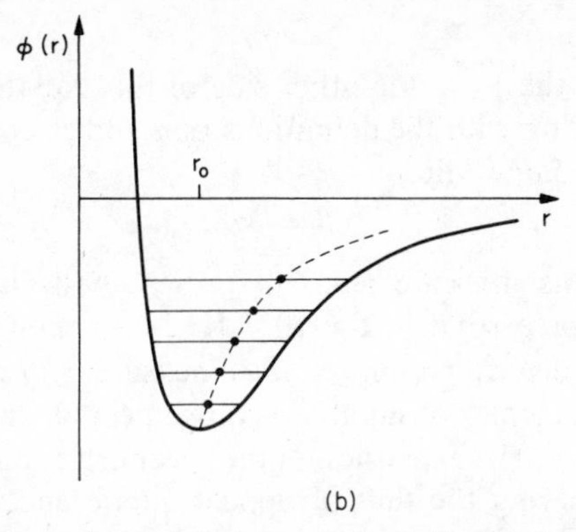

Figure 5.5 The origin of thermal expansion. As the temperature is raised, the atoms vibrate with increasing amplitude about their equilibrium position. The horizontal lines represent various total energies of vibration. r will oscillate between the values it takes where the horizontal lines cut the potential energy curves. The mean interatomic separation at each level of oscillation is represented by the dots. If the potential is symmetric about the minimum (a), the mean separation is unchanged; but if it is asymmetric as shown in (b) the mean (time average) separation increases as the amplitude of vibration increases.

The units of α are K^{-1}. In differential notation, α is given by

$$\alpha = \frac{1}{L}\left(\frac{\partial L}{\partial T}\right)_F$$

The suffix F reminds us that it is assumed in the definition of α that any tensional force is kept constant as the material expands.

One may also define a *cubic expansivity* β* which is the fractional rate of increase of volume with temperature:

fractional increase in volume = β × increase in temperature

or, in symbols,

$$\frac{\Delta V}{V} = \beta \, \Delta T \qquad (5.6)$$

In differential notation, β is given by

$$\beta = \frac{1}{V}\left(\frac{\partial V}{\partial T}\right)_p$$

The suffix p reminds us that it is assumed in the definition of β that any pressure on the material is kept constant during the expansion.

The two expansivities are normally related in a very simple way. Suppose we have a cube of the material of side L_0, and that the material is *isotropic* (that is, that it behaves the same way in all directions). Then if it expands so that each side of the cube increases in length by ΔL, the new volume will be

$$V = (L_0 + \Delta L)^3$$

$$\approx L_0^3\left(1 + 3\frac{\Delta L}{L}\right)$$

* β is the Greek letter *beta*.

where we have expanded the bracket with a binomial expansion and only taken the first two terms. (The remaining two terms are second and third order small quantities—the square and cube of $\Delta L/L$ which is a small number—and are therefore much smaller than the terms we retain.) But the original volume is $V_0 = L_0^3$, so the change in volume is

$$\Delta V = 3V_0\frac{\Delta L}{L}$$

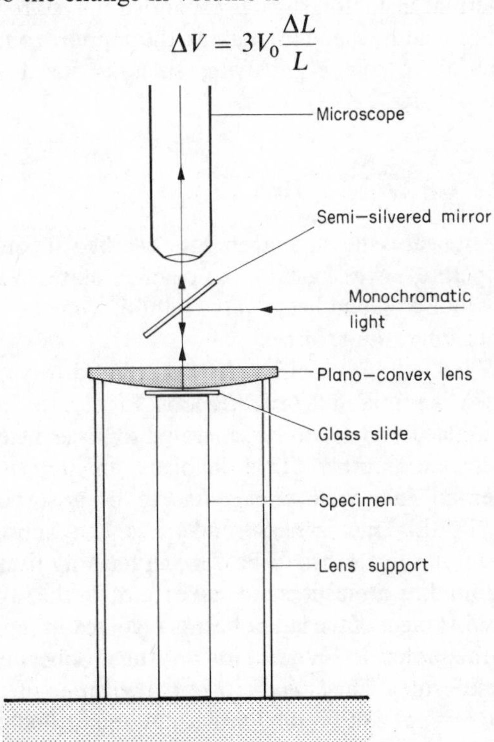

Figure 5.6 An arrangement for measurement of linear expansivity. The expansion of the specimen is measured by observing the shift of interference fringes formed between the glass slide and the lower surface of the lens (Newton's rings).

Taking the V_0 to the other side of this equation and comparing with the definitions of α and β, equations *5.4* and *5.6*, we find

$$\beta = 3\alpha \qquad (5.7)$$

Expansivities are generally rather small—a typical value for a solid is $\alpha \approx 10^{-5}\,\mathrm{K}^{-1}$—so their experimental determination requires measurement of small displacements. A suitable arrangement is shown in figure 5.6. The expansion of the specimen is measured by observing the shift of optical interference fringes formed between the lower face of the plano-convex lens and the upper surface of a glass microscope cover slide which rests on the specimen (Newton's rings). The specimen may be heated electrically and its temperature measured with a thermocouple or by inserting a mercury in glass thermometer into it. The lens and specimen must be supported firmly to give adequate mechanical stability, but care must be taken to avoid any significant expansion of the lens supports as the specimen is warmed. Movement of the lens may be reduced by (a) surrounding the specimen with a thermal insulator, such as expanded polystyrene, to reduce heat loss, and (b) making the supports out of a material of low expansivity, such as fused silica ($\alpha \approx 4 \times 10^{-7}\,\mathrm{K}^{-1}$).

5.4 LATENT HEATS

Latent heat is the thermal energy absorbed at constant temperature when a material changes state. We will now look at latent heats in the light of what we know about interatomic forces.

When a substance changes from liquid to vapour, energy is required for two reasons. Firstly, the atoms (or molecules) have to be separated to large distances from one another. This involves an increase in potential energy, so energy has to be absorbed to supply this increase. Secondly, as the substance evaporates, work has to be done in pushing back the surrounding atmosphere to make room for the vapour. Usually, most of the latent heat is involved in separating the molecules. We will look at the numbers in the case of water. The *latent heat of vaporization* of water at 100 °C is $4.06 \times 10^4\,\mathrm{J\,mol^{-1}}$. We may find what proportion of this is involved in pushing back the atmosphere by working from the molar volume. The molar volume of an ideal gas at s.t.p. is $2.24 \times 10^{-2}\,\mathrm{m}^3$. The molar volume of water vapour at 100 °C and 1 atm pressure is therefore about

$$(2.24 \times 10^{-2} \times \tfrac{373}{273})\,\mathrm{m}^3 = 3.06 \times 10^{-2}\,\mathrm{m}^3$$

where we have assumed that water vapour does not behave too differently from an ideal gas. The work done in pushing back the atmosphere by this volume is

$$(1\text{ atm}) \times (\text{volume}) = 1.01 \times 10^5 \times 3.06 \times 10^{-2}\,\mathrm{J}$$
$$= 3.09 \times 10^3\,\mathrm{J}$$

Therefore only about 7.6% of the total energy is involved in pushing back the atmosphere. The rest is used to separate the molecules.

Can we relate the latent heat of vaporization to the potential energy of two molecules at their equilibrium separation, ε of figure 5.2? In the liquid, the distance between a molecule and its neighbours will be about r_0, so the potential energy associated with each 'bond' to neighbours will be about ε. Figure 5.2 shows that the potential energy to atoms beyond the near neighbours ($r \gtrsim 2r_0$) will be much smaller than to the near neighbours and so may be neglected. We may therefore think of a liquid as consisting of atoms or molecules, close together but not regularly arranged, bound to their near neighbours only (figure 5.7). Suppose there are z near neighbours. This is called the *coordination number*. Then the number of bonds in a mole of material is $zN_A/2$, the factor $\frac{1}{2}$ being there because each bond has a molecule at each end (i.e. there are two molecules per bond). Then the total energy to separate all the molecules in a mole is $zN_A\varepsilon/2$. Since work done against external pressure is usually small, this will be approximately equal to the molar latent heat of vaporization, L_m:

$$L_m \approx \tfrac{1}{2}zN_A\varepsilon \qquad (5.8)$$

Typically, $z \approx 10$ so we can use measured latent heats of vaporization to estimate ε. Substituting $L_m =$

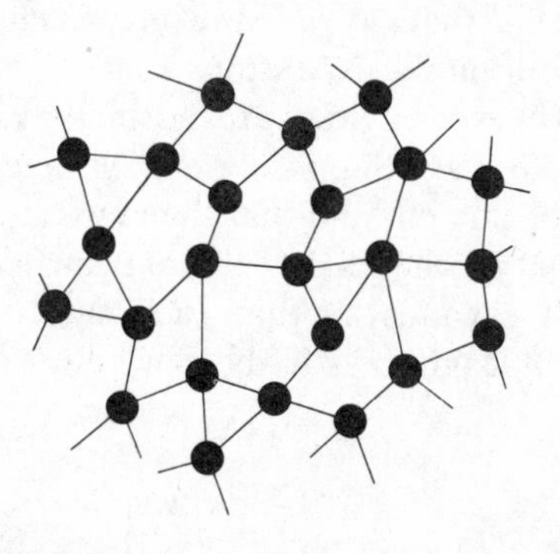

Figure 5.7 A liquid may be thought of as consisting of atoms or molecules close together but not regularly arranged and bonded to their near neighbours

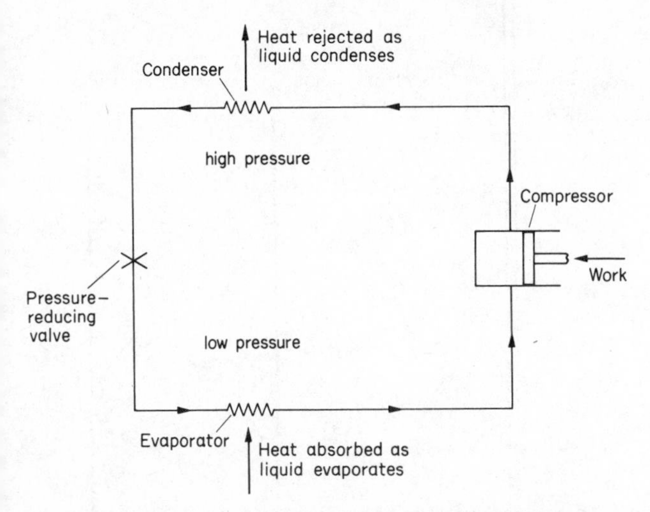

Figure 5.8 The elements of the common domestic refrigerator

4.06×10^4 J mol^{-1} for water (neglecting the proportion used in pushing back the atmosphere) we get

$$\varepsilon \approx 1.3 \times 10^{-20}\ \text{J} \approx 84\ \text{meV}$$

which is much weaker than a chemical bond but stronger than a typical van der Waals' bond between two atoms (table 5.1, page 75).

Latent heat of vaporization is used in the common domestic refrigerator (figure 5.8). The *working fluid* is a volatile liquid which is circulated by an electrically driven compressor. The compressor takes in vapour from the low pressure part of the circuit and compresses it to above the saturated vapour pressure (page 81) so that it condenses, giving up its latent heat which is rejected through a heat exchanger to the surrounding atmosphere. (The exchanger with its cooling fins is usually visible at the back of a refrigerator.) The liquid is forced through a valve into the low pressure part of the circuit where, the pressure being low, it boils, absorbing its latent heat through another heat exchanger inside the refrigerator. The refrigerator is a *heat pump* (section 7.7), a device which uses energy to pump heat in the unnatural direction from cold to hot.

When a solid melts, there is usually very little change of volume so the work against the atmosphere is negligible. In the solid the molecules are regularly arranged in fixed positions, but in the liquid they are mobile. For the molecules to have room to move around in the liquid, their mean separation must usually be a little greater than in the solid. Also, they must have spaces to move into, so the number of near neighbours will decrease slightly. The latent heat of melting supplies the energy needed to (a) adjust the intermolecular distances, and (b) reduce the effective number of bonds; but the molecules stay close together, so we would expect the *latent heat of fusion* to be much smaller than the latent heat of vaporization where all the bonds are broken. This is exactly what we find. The latent heat of melting of ice, for example, is 6.0×10^3 J mol^{-1}, only about 15% of the latent heat of vaporization.

5.5 SURFACE TENSION

Surface tension is a measure of the work that has to be done to create new surface. The existence of surface tension means that it costs energy to move an atom or molecule from inside a liquid to the surface. This is easy to understand in terms of the attraction between a molecule and its neighbours in the liquid. Within the liquid a molecule is surrounded by neighbours on all sides. Typically, there are about 10 near neighbours. When a molecule is in the surface, however, it has neighbours on one side towards the bulk of the liquid, but none on the other. It therefore loses about half its bonds when moved into the surface. Now the bonding energy per molecule is L_m/N_A where L_m is the molar latent heat of vaporization, so the energy required to move one molecule into the surface will be about $L_m/2N_A$. If $\mathcal{N}$ is the number of molecules per area of surface, the work required to produce unit area of new surface will be about $\mathcal{N} L_m/2N_A$. Therefore we would expect the surface tension to be given by

$$\gamma \approx \mathcal{N} L_m/2N_A \approx z\mathcal{N}\varepsilon/4 \qquad (5.9)$$

where we have used *5.8* to substitute for L_m.

We can use this formula to give another estimate of ε. For water, $\gamma \approx 7.3 \times 10^{-2}$ N m^{-1}. We may estimate $\mathcal{N}$ by supposing the molecules to be arranged in regular, equally spaced rows along three mutually perpendicular directions: a cubic array (figure 5.9). (They are not arranged regularly in a liquid, of course, but this must give numbers which are roughly correct.) If there are n molecules per unit length in each direction, the number per unit area is

$$n^2 = \mathcal{N}$$

and the number per unit volume is

$$n^3 = N_A/V_m$$

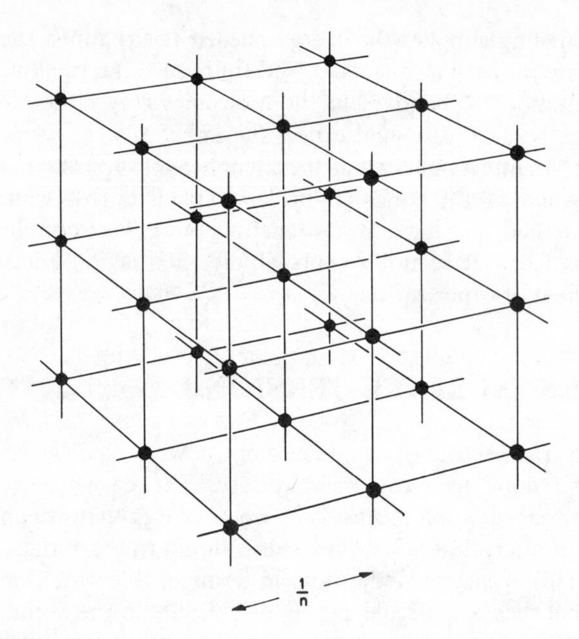

Figure 5.9 A cubic array of atoms. If there are n atoms per unit length along the rows, there are n^2 per unit area and n^3 per unit volume.

Hence, eliminating n,

$$\mathcal{N} = (N_A/V_m)^{\frac{2}{3}}$$

The molar volume of water ($M_r = 18$ and density $= 10^3$ kg m^{-3}) is 1.8×10^{-5} m^3 (= the volume of (M_r/1000) kg, see page 46). Hence, $\mathcal{N} = 1.04 \times 10^{19}$ m^{-2}. Substituting in *5.9* and putting, $z = 10$, we get $\varepsilon \approx 2.8 \times 10^{-21}$ J = 18 meV which is a quarter of the value estimated from latent heat (not bad agreement for calculations based on such simple models).

A common way of determining surface tension is to measure the rise of the liquid in a thin capillary (figure 5.10). If the radius of the capillary is r, and the liquid meets the solid surface at an angle θ (called the *angle of contact*), the radius of the meniscus is $r/\cos\theta$. Then using equation *2.13* (page 23) for the pressure difference across a spherical surface due to surface tension, we have a pressure difference

$$\Delta p = 2\gamma \cos\theta/r$$

This pressure difference must just support the weight of the liquid below, so

$$\rho g h = 2\gamma \cos\theta/r \qquad (5.10)$$

where ρ is the liquid density. Note that if $\theta > 90°$, the liquid is *depressed* in the capillary.

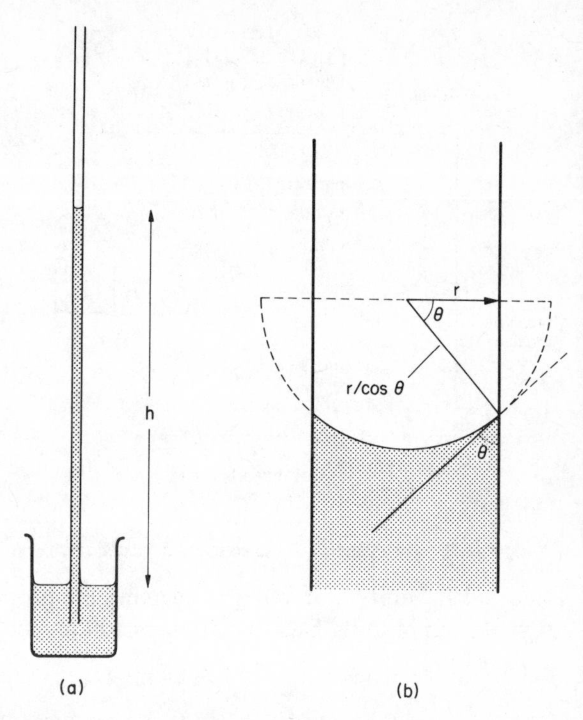

Figure 5.10 Measurement of surface tension by capillary rise. (a) capillary rise; (b) the geometry at the meniscus

What determines the angle of contact? If there is a strong attraction between the tube and molecules of the liquid, the surface of the tube will rapidly become covered with a monomolecular layer (a layer one molecule thick) of the liquid by absorbing molecules from the vapour. The liquid will readily join on to this surface layer and the angle of contact will be zero: the liquid *wets* the surface. This is the case with contact between water and clean glass. If bonding between the surface and liquid is negligible, there will be no tendency for the liquid to wet the tube. Instead, the liquid will simply try to reduce its surface area (and hence its surface energy) by taking a convex shape at the meniscus corresponding to $\theta = 180°$. Bonding between mercury and glass is very weak, and for them, $\theta \approx 137°$.

Where does the energy for capillary rise come from? It comes from intermolecular potential energy. If the surface and liquid have lower potential energy when they are in contact because of large attractions between the tube and molecules of the liquid, then the energy is higher when they are not in contact. When the liquid climbs the tube, this potential energy falls as the area of contact between liquid and tube increases. The energy released is available to make up for the increase of gravitational potential as the liquid rises.

5.6 VAPOUR PRESSURE

Consider a liquid and its vapour inside an otherwise evacuated container. The potential energy of a molecule is lower in the liquid than in the vapour because, in the liquid, it is close to other molecules to which it is attracted. But this does not mean that *all* the molecules will be in the liquid because there will always be some which have enough thermal energy to escape from the liquid into the vapour. When the liquid and vapour are in equilibrium, the rate of escape will equal the rate at which molecules recondense from the vapour into the liquid: the equilibrium is *dynamic*. When in equilibrium with the liquid, the vapour is said to be *saturated* and the pressure it exerts is the *saturated vapour pressure*. We shall now see what we can deduce about the saturated vapour pressure of a liquid.

According to the Boltzmann factor (page 53), the ratio of the probability that a molecule is in a vapour state to the probability that it is in a liquid state is $\exp -(\Delta\phi/kT)$ where $\Delta\phi$ is the increase in energy on taking a molecule from the liquid into the vapour. But the molar latent heat of vaporization L_m is the energy required to take N_A molecules from liquid to vapour. Therefore, $\Delta\phi = L_m/N_A$, and the density of molecules in the vapour n will be related to temperature by

$$n \propto \exp -(L_m/N_A kT) = \exp -(L_m/RT)$$

But, the pressure $p = nkT$ (page 47), so

$$p \propto T \exp -(L_m/RT) \qquad (5.11)$$

We may test this equation by using the variation of vapour pressure with temperature to calculate a value for L_m. Data for water are:

$$p = 9.2 \text{ mmHg} \quad \text{at} \quad 10\,°\text{C}$$
$$p = 17.5 \text{ mmHg} \quad \text{at} \quad 20\,°\text{C}$$

Taking logs of *5.11*

$$\ln(p/T) = -(L_m/RT) + \text{constant}$$

Substituting both sets of data and subtracting the resulting equations to eliminate the constant

$$\ln(p_2/T_2) - \ln(p_1/T_1) = (L_m/RT)(1/T_1 - 1/T_2)$$

Substituting the values for water we obtain*

$$L_m = 4.2 \times 10^4 \text{ J mol}^{-1}$$

* Note that it is not necessary to change the units of pressure because any conversion factor cancels out: the left hand side of the equation is $\ln(p_2T_1/p_1T_2)$, so it does not matter what the units of p_1 and p_2 are (as long as they are the same).

which agrees well with the measured value at 15 °C: 4.43×10^4 J mol^{-1}. (If more data were given, one should plot a graph of $\ln(p/T)$ against $1/T$ which would be a straight line of gradient $-L_m/R$.)

A more detailed analysis of vapour pressure gives a formula like *5.11* but with $T^{3/2}$ in front of the exponential. However, the variation of vapour pressure with temperature is dominated by the exponential and changing the power of the T in front makes very little difference to the form of the dependence. In fact, the preexponential temperature factor is so unimportant that, to a good approximation, it may be neglected entirely, putting

$$p \propto \exp -(L_m/RT) \qquad (5.12)$$

With this approximation, a plot of $\ln p$ against $1/T$ is a straight line of gradient $-L_m/R$. (See problem 5.27.)

A simple way of measuring vapour pressure is to introduce a small amount of the liquid above the mercury in a barometer (figure 5.11). With no liquid present, the height of the mercury column corresponds to atmospheric pressure. With liquid present, the pressure above the mercury is the saturated vapour

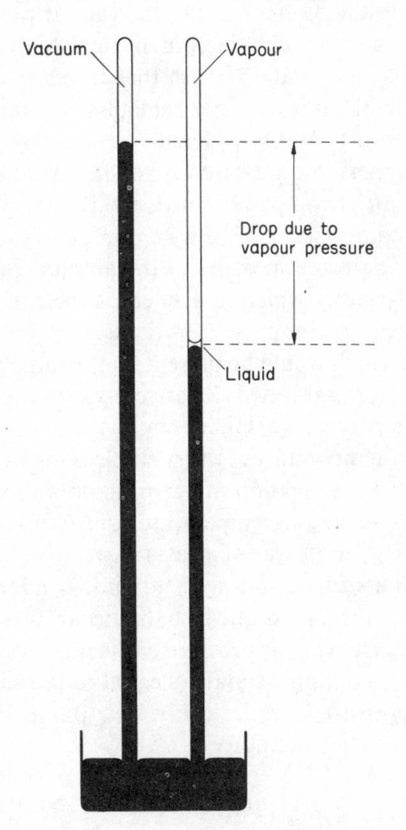

Figure 5.11 Measurement of vapour pressure

pressure of the liquid so the level falls by an equivalent amount. For measurements at different temperatures, the barometer tube can be surrounded by a heating or cooling jacket.

The pressure of a vapour can, of course, be less than the saturated vapour pressure of the liquid at the temperature concerned. If, for example, we had introduced so little liquid into the barometer tube of figure 5.11 that it all evaporated, the pressure would simply have been determined by the gas law for the amount of vapour present. The vapour would then be said to be *unsaturated*; it would not be in equilibrium with liquid if liquid were present. If more liquid were introduced into the tube, some would evaporate to bring the pressure up to the saturated value. In contrast, vapour pressure cannot normally rise *above* the saturated value because liquid will usually condense out. (But see section 5.7.)

If we have a vapour mixed with other gases, the total pressure will be given by the sum of the partial pressures of each gas present (page 49). In the simple method for measuring vapour pressure outlined above, the mercury itself will contribute to the total pressure in the barometer, but the vapour pressure of mercury is extremely small at normal temperatures (7×10^{-4} mmHg at 15 °C) so that its effect is normally negligible. If it were not negligible, would the *fall* in the mercury level be different?

Atmospheric air is a case where the partial pressure of a vapour is comparable with the partial pressures of other components. Air always contains some water vapour. The usual measure of the amount present is *relative humidity* which is defined as the ratio of the mass of water vapour present (in a given volume) to the mass which would be present (in the same volume) if the air were saturated. Relative humidity is usually given as a percentage. High humidity when combined with high temperatures is very unpleasant because it inhibits the evaporation of perspiration and so interferes with the body's temperature control mechanism.

A fairly accurate way of measuring relative humidity is to cool a surface until moisture just condenses out on it. The temperature at which this occurs is called the *dew point*. The vapour pressure of the water present in the air is then (approximately) equal to the saturated vapour pressure of water at the dew point. We then have for relative humidity,

$$\text{r.h.} = \frac{\text{mass present}}{\text{mass if saturated}}$$

$$\approx \frac{\text{saturated vapour pressure of water at the dew point}}{\text{saturated vapour pressure of water at the ambient temperature}}$$

The dew point is so called because the same mechanism causes formation of dew at night: on clear nights, the earth's surface loses heat by radiation into space and it therefore drops in temperature. When the temperature of exposed surfaces falls to the dew point, water condenses out. Frost forms in the same way except that condensation occurs directly from vapour to solid when the surface temperature is below freezing point. Formation of both dew and frost is inhibited by wind which keeps the air stirred so preventing the surface temperature from falling so much.

5.6.1 When a metal is heated to a high temperature, it emits electrons. This is known as *thermionic emission* and is used as the source of electrons in cathode ray tubes and radio valves. The process of emission is equivalent to evaporation of molecules from a liquid: the metal acts as a potential 'well' in which the electrons are normally trapped; but unusually energetic ones have enough kinetic energy to escape over the potential barrier and 'evaporate' from the metal surface (figure 5.12). The height of the potential barrier ϕ_0 is called the *work function*. A typical value is 4 eV, so the Boltzmann factor giving the proportion of electrons which are energetic enough to escape is very small: at 1000 K it is

$$\exp -(4 \times 1.6 \times 10^{-19}/1.38 \times 10^{-23} \times 10^3)$$
$$= \exp -46.6 = 1.4 \times 10^{-20}$$

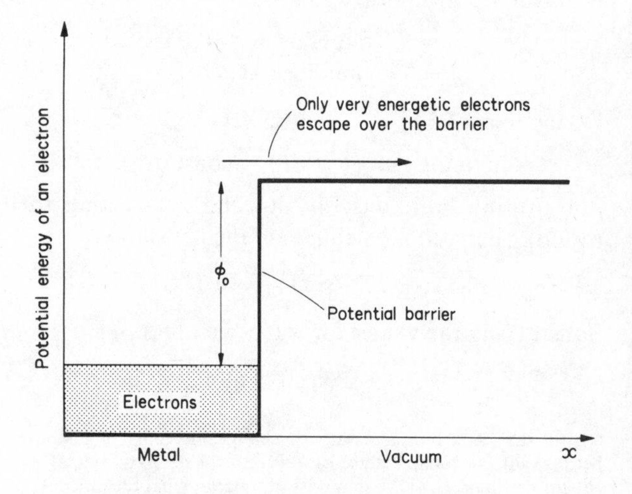

Figure 5.12 The energetics of thermionic emission

(The factor 1.6×10^{-19} converts eV to joules.) Nevertheless, metals contain such a high density of electrons that quite large currents can be emitted.

5.7 VAPOUR PRESSURE OVER CURVED SURFACES

To evaporate from the liquid into the vapour, a molecule has to escape from the attraction of the neighbouring molecules in the liquid. If the surface of the liquid is curved, the number of molecules holding it back into the liquid is changed so the vapour pressure changes. This is illustrated in figure 5.13.

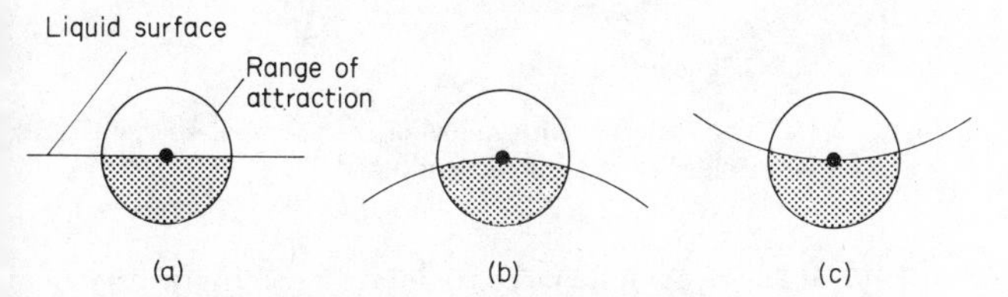

Figure 5.13 Qualitative explanation of the variation of vapour pressure with surface curvature. The shaded regions represent the volumes where other bonding molecules are likely to be present. Compared with the case of a plane surface (a), the number of bonds is likely to decrease when the surface is convex (b), and increase when it is concave (c).

The dot represents the molecule which is trying to escape and the sphere around it shows the range of the attractive forces: the molecule is attracted strongly to others within the sphere, but forces to those outside are small. The forces fall off very quickly with distance (see figure 5.1, page 73) so the range is not much more than a molecular diameter. If the surface of the liquid is flat, the molecule will be bound to any others within the lower hemisphere (shaded in (a)). If the surface is convex (b), the volume in which neighbours will be present is reduced, so the binding to the liquid is reduced and the vapour pressure rises. Conversely, if the surface is concave (c), the molecule is likely to be bound to the liquid by more 'bonds' and the vapour pressure will drop. Because the range of the forces is so small, the effect will only become important for very curved surfaces. We shall now deduce a formula for how the vapour pressure is affected by the following simple argument.

Consider a vertical tube dipping into a beaker of the liquid inside an enclosure which is evacuated apart from vapour of the liquid (figure 5.14). Because of surface tension, the liquid will rise in the tube. The rise h will be given by

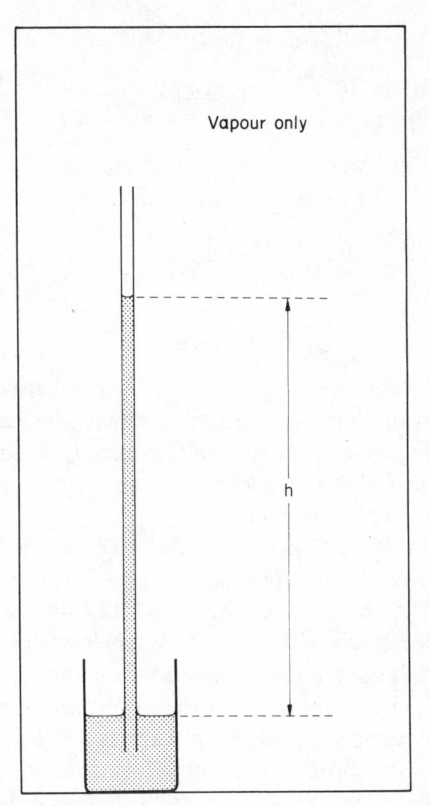

Figure 5.14 Derivation of the dependence of vapour pressure on surface curvature

$$h\rho_l g = 2\gamma/r$$

where ρ_l is the density of the liquid, r the radius of curvature of the *meniscus* (see equation *5.10*), and γ the surface tension. Now when the liquid has risen in the tube, the whole system must reach an equilibrium state. In particular, the vapour just above the flat liquid surface in the beaker must be in equilibrium with the liquid there, and the vapour just above the meniscus in the tube must be in equilibrium with the liquid there. (If this were not so, we would end up with a steady circulation and could build a perpetual motion machine.) But the vapour pressure at the height of the meniscus is less than it is at the flat surface because the vapour is in a gravitational field. The drop in vapour pressure is

$$\Delta p = h\rho_v g = 2\gamma\rho_v/r\rho_l \qquad (5.13)$$

where ρ_v is the density of the vapour and we have assumed the effect to be relatively small so that ρ_v is approximately constant over the height h. If we approximate the vapour to an ideal gas, we may use the perfect gas law for the vapour pressure p_0,

$$RT/M = p_0V_m/M = p_0/\rho_v$$

where M is the molar mass. Substituting in *5.13*, we get a formula for the fractional change in vapour pressure:

$$\Delta p/p_0 = 2\gamma M/r\rho_l RT \qquad (5.14)$$

Substituting for water with $\gamma = 7.3 \times 10^{-2}\,\mathrm{N\,m^{-1}}$, $M = 1.8 \times 10^{-2}$ kg, $\rho_l = 10^3\,\mathrm{kg\,m^{-3}}$ and $T = 300$ K, we get

$$\Delta p/p_0 = 1/(r/\mathrm{nm}) \qquad (5.15)$$

In equations *5.14* and *5.15* we have taken the sign convention that r is positive for a convex surface so the vapour pressure increases (Δp positive), and r is negative for a concave surface so vapour pressure decreases (Δp negative).

Equation *5.15* shows that the effect is small except for very small drops (r positive) or bubbles (r negative). For a 1% change, the radius must be one sixth of the wavelength of visible light! Nevertheless, the dependence of vapour pressure on surface curvature is very important in connection with the formation of droplets by condensation from vapour. Formation of bubbles in a liquid turns out rather differently, as we shall see.

Consider what happens when a droplet is immersed in its vapour. When the droplet is very small, r is small (and positive) so the vapour pressure just outside the droplet is larger than it would be over a plane surface of the liquid at the same temperature. For the droplet to be in equilibrium with the surrounding vapour, we must have

$$p_v = p_0 + 2\gamma\rho_v/r\rho_l$$

where p_v is the pressure of the surrounding vapour and p_0 is the saturated vapour pressure (above a plane surface) of the liquid. But if the droplet then becomes a little smaller, its vapour pressure will rise above the pressure of the surrounding vapour and it will evaporate; while, if it becomes a little bigger, its vapour pressure decreases below the pressure of the surrounding vapour which then condenses on the droplet and makes it grow still more. The equilibrium is therefore *unstable*. It corresponds to a *maximum* in the potential energy as a function of drop radius (figure 5.15). Furthermore, the vapour pressure of the drop continues to increase as its radius decreases, so how does it ever start to form in the first place? The answer is that there are usually small particles in the vapour on which vapour can condense without the

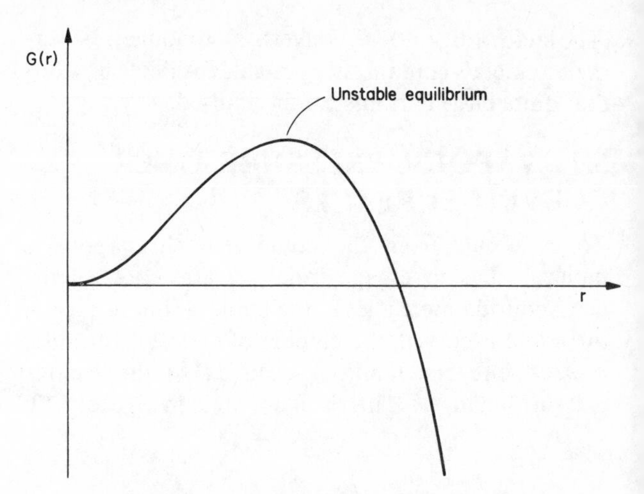

Figure 5.15 The variation with radius of the potential energy *G* of a droplet in a vapour or of a bubble in a liquid

liquid surface ever having to have a small radius of curvature. The result is that droplets will normally form without the pressure of the vapour having to rise significantly above the value it would have over a plane liquid surface (i.e. the saturated vapour pressure). But if no *condensation nuclei* are present, the vapour can be *supercooled* (cooled to a temperature at which the pressure of the vapour is greater than would be in equilibrium with a plane liquid surface) without condensation taking place. In the atmosphere, droplets condense out to form cloud at *supersaturations* of only a few per cent. This is because dust and active chemical complexes are present to act a condensation nuclei.

Supercooling is used in the *Wilson cloud chamber*, a device for making visible the tracks of charged particles emitted in radioactivity. Clean air, saturated with water vapour is contained in a cylinder closed at the top with a glass plate and at the bottom by a piston (figure 5.16). The piston is lowered quickly so that the gas makes an adiabatic expansion and cools. Because it cools, the vapour becomes supersaturated, but condensation does not take place generally through the volume of gas because there are no nuclei. Only where fast charged particles have passed, leaving a trail of ions to act as condensation nuclei, do droplets form. The droplets show up clearly when the chamber is illuminated so the tracks of the particles are made 'visible' (figure 5.17).

Nucleation is also of great importance in connection with the *boiling* of liquids. (The difference between evaporation and boiling is that in evaporation the vapour is lost from the surface of the liquid,

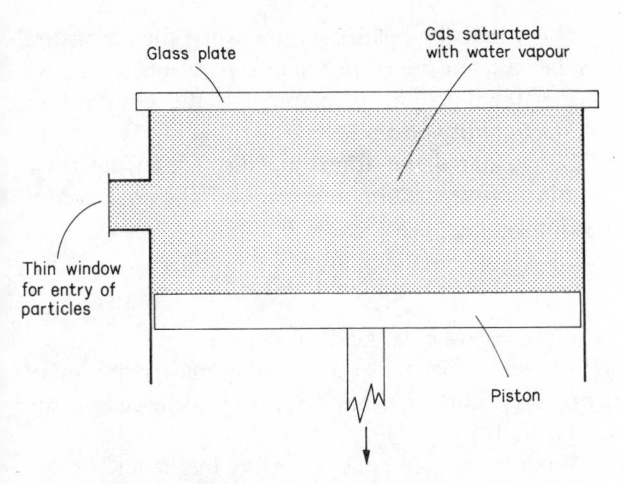

Figure 5.16 The elements of a Wilson cloud chamber

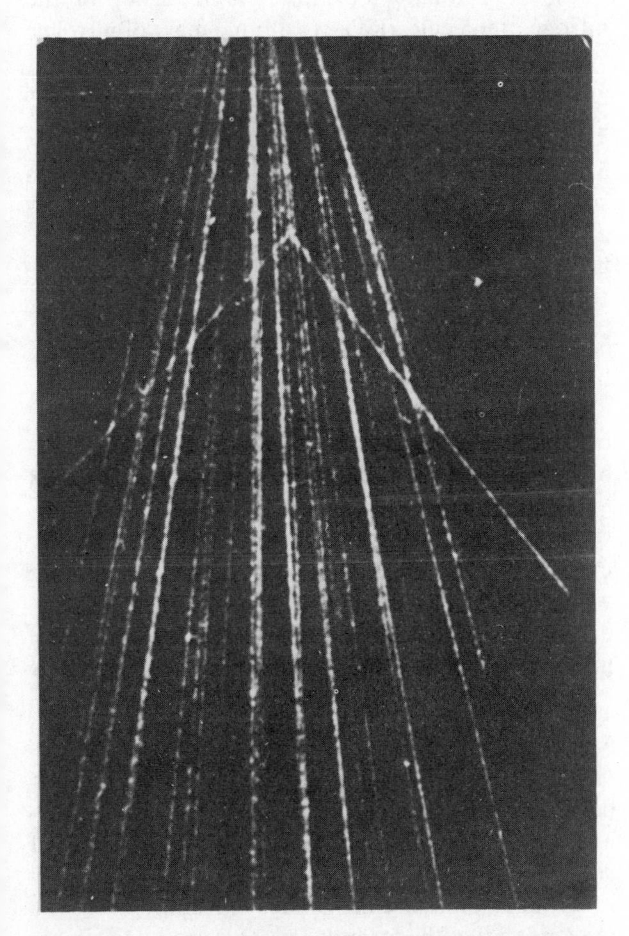

Figure 5.17 A photograph of tracks in a Wilson cloud chamber. The chamber was filled with helium and alpha particles (nuclei of helium atoms) enter from above. The forked track results from a collision between an alpha particle and a helium atom of the gas, both particles leaving tracks after the collision. The tracks are nearly at right angles. Do you know why?

whereas in boiling, the vapour pressure is so high that bubbles of vapour form inside the liquid. Usually they go on growing as they rise to the surface.) There is, however, a point of contrast: in the case of boiling, the need for nucleation is primarily the result of the *pressure difference* across the curved liquid surface around the bubble. The effect of the change of vapour pressure is relatively small, as we shall now show.

If the pressure in the liquid is p_l, the pressure inside a small bubble of radius r is $p_l + 2\gamma/r$ where γ is the surface tension of the liquid. The vapour pressure of the liquid in the surface surrounding the bubble is decreased as a result of the curvature to $p_0 - 2\gamma\rho_v/r\rho_l$, where p_0 is the saturated vapour pressure (over a plane surface). For equilibrium between the vapour enclosed in the bubble and the surrounding liquid we therefore need

$$p_0 - 2\gamma\rho_v/r\rho_l = p_l + 2\gamma/r$$

Again, equilibrium is unstable: if the radius increases, the pressure of the vapour inside the bubble decreases and the vapour pressure of the surrounding liquid increases so that the liquid evaporates into the bubble and the bubble continues to grow. Conversely, if the radius decreases, the pressure of the vapour in the bubble becomes greater than the vapour pressure of the surrounding liquid so that the vapour condenses on the liquid and the bubble collapses. The potential energy again varies with radius as shown in figure 5.15. The condition for equilibrium may also be written

$$p_0 = p_l + \frac{2\gamma}{r}\left(1 + \frac{\rho_v}{\rho_l}\right)$$

The two terms in the bracket come from the two ways in which the equilibrium is affected by the change in drop radius: the first comes from the pressure change across the curved surface and the second from the change of vapour pressure. Since $\rho_v/\rho_l \ll 1$ (typically, $\rho_v/\rho_l \approx 1/1000$), it is obvious that the change of pressure across the curved surface is by far the more important effect.

Thus, if a liquid is to boil without the temperature rising significantly above the value at which the vapour pressure over a flat surface is atmospheric (the normal boiling point), nuclei are again necessary. If a liquid is very pure, however, so that no nuclei are present, and it is heated in a smooth-walled container, it may become *superheated* (heated above its normal boiling point) so that when bubbles do eventually form, they grow so violently that the liquid 'bumps'. This can be avoided by adding to the liquid pieces of some inert rough material, such as broken earthenware, to provide a suitable surface for bubble nucleation.

Nucleation of bubbles in a superheated liquid is used in the investigation of nuclear interactions in *bubble chambers*. These operate in a way similar to the Wilson cloud chamber except that the chamber contains a liquid close to its boiling point. The pressure applied to the liquid is reduced suddenly to below the saturated vapour pressure so that the liquid becomes superheated. Charged particles passing through the liquid produce local heating which nucleates bubble formation so that the tracks are made visible as a trail of small bubbles. An important difference between bubble and cloud chambers is that the matter in the bubble chamber is much more dense (a liquid instead of a gas) so that there is a much greater chance of incident particles making collisions within the chamber. This makes the bubble chamber a much more efficient device for the study of nuclear interactions than the cloud chamber.

PROBLEMS

Interatomic force and potential energy; size of atoms and molecules. Origins of interatomic forces; bond energies.

5.1 What is meant by the 'size' of an atom or molecule? How may it be measured? Explain why the apparent size of an atom depends on what experiment is used to measure it.

5.2 The density of benzene is 880 kg m^{-3} and its relative molecular mass 78.1.
a) What is the volume occupied by one benzene molecule?
b) Estimate the radius of a spherical molecule which would take up the same space.

Elasticity: atomic origin; Hooke's law; Young modulus and bulk modulus

5.3 What is the relationship between potential energy and force?
Two identical atoms, each of mass m, are bound together to form a molecule. Close to the equilibrium separation r_0, the interatomic potential energy $\phi(r)$ increases as the square of the displacement:

$$\phi(r) = \phi(r_0) + a(r - r_0)^2$$

a) Show that the restoring force when the atoms are displaced from the equilibrium separation is directly proportional to the displacement, the constant of proportionality being $2a$.
b) Show that if the atoms are displaced from their equilibrium separation and released, the equation of motion is

$$m\ddot{x} + ax = 0$$

where $x = r - r_0$ is the displacement.
(*Hint*: remember that the centre of mass will remain stationary since there are no external forces on the molecule.)
c) What is the molecular vibration frequency?
d) For a molecule like N_2, the constant a is of order 30 J m^{-2}. Estimate the vibration frequency of the nitrogen molecule. (Relative atomic mass of nitrogen = 14.)
e) To approximately what temperature does one quantum of oscillation correspond?

5.4 A rod 10 m long and 10 mm in diameter is stretched 1.9 mm when supporting a load of 1500 kg.
a) What is the stress?
b) What is the strain?
c) What is the Young modulus?

5.5 A wire of unstretched length L and density ρ hangs from one end.
a) Find the stress in the wire a distance x from the bottom end.
b) Hence, show that a short element of the wire dx long and x from the bottom extends by $\rho g x\,\mathrm{d}x/E$ where E is the Young modulus.
c) Hence show that the total increase in length of the wire is $\rho g L^2/2E$.
d) Explain in words why, in this situation, the increase in length is not directly proportional to the original length.

5.6 The pendulum of a grandfather clock consists of a 1 kg mass suspended on a steel rod of length 0.9 m and cross sectional area 10 mm^2.
a) How much does the rod stretch as a result of supporting the mass?
b) What is the fractional change in length?
c) Write down the formula for the period of a simple pendulum.
d) What is the fractional change in the period of the pendulum as a result of the stretching of the rod?
[For steel, $E = 2 \times 10^{11}$ Pa.]

5.7 5×10^{-3} m^3 of water are contained in a cylinder with piston. If the compressibility of water is 4×10^{-10} Pa^{-1}, how much work is done in increasing the pressure isothermally from 1 to 250 atm?

Thermal expansion; molecular origin; linear and cubic expansivities

5.8 How serious are changes in the timekeeping of a clock resulting from thermal expansion in the pendulum?
[For a metal, α is of order 1×10^{-5} K^{-1}.]

5.9 A steel rod of radius 5.00 mm is to be fixed in a hole in a steel block by making the hole slightly too small, heating the block until it expands enough for the rod to be inserted, and allowing the block to shrink onto the rod as it cools. If the hole is machined to a radius of 4.99 mm, by how much must the block be heated before the rod can be inserted?
[For steel, $\alpha = 1.1 \times 10^{-5}$ K^{-1}.]

5.10 A steel wire, 3 m long, supports a load of 10 kg. The temperature is raised by 50 °C. How much work is done on the load?
[Linear expansivity of steel = 1.1×10^{-5} K^{-1}.]

5.11 An experimental arrangement like that shown in figure 5.6 is used to measure the linear expansivity of brass. As the specimen is gradually warmed, the Newton's rings expand, and each time the pattern passes through its original position the temperature t is recorded:

t/°C: 15.1 15.9 16.7 17.4 18.2 19.0 19.8 20.5

The fringes are formed with sodium light and the specimen, which is 20 mm long, is well lagged so that heating of the supports is negligible.
a) By how much does the thickness of the film of air between the slide and the lens have to change for the ring pattern to reach its original position (i.e. for the pattern to move by one fringe)?
b) What is the linear expansivity of brass?
c) What accuracy would you quote with your result?
d) How would you improve the accuracy of the experiment?
(Note the sensitivity which results from using light waves to measure the change in length.)
[Mean wavelength of sodium D lines = 589.3 nm.]

5.12 What is the cubic expansivity of an ideal gas at temperature T?

5.13 A clinical thermometer is to use 60 mm^3 of mercury and must have the 1 °C marks 1 cm apart. What size bore must be used for the capillary? (Neglect expansion of the glass.)
[Volume expansivity of mercury = 1.82×10^{-4} K^{-1}.]

5.14 A burette is calibrated for 15 °C. It is used to measure out known masses of a solution whose density at 15 °C is known. If the measurements are actually made at 20 °C, and no correction is made for the difference of temperature,
a) what is the percentage error in the volumes actually measured?
b) what is the percentage error in the calculated masses that are thought to be measured out?
c) are the calculated masses too large or too small?
d) does the expansion of the liquid or of the glass contribute most to the error?
[Linear expansivity of the glass, $\alpha_g = 8 \times 10^{-6}$ K^{-1}.
Cubic expansivity of the liquid, $\beta_l = 2 \times 10^{-4}$ K^{-1}.]

5.15 A brass rod of 10 mm^2 cross sectional area is rigidly clamped at both ends. The temperature falls by 10 °C. What is the tension in the rod?
(*Hint:* Find how much a rod of length L would contract as a result of the cooling if it were not clamped at the ends, and then calculate the tension which would stretch it to its original length.)
[Linear expansivity of brass = 1.9×10^{-5} K^{-1}.
Young modulus of brass = 1.0×10^{11} Pa.]

5.16 100 ml of water completely fill a strong steel container. The temperature is raised by 10 K.
a) By how much would the water increase in volume if it were free to expand?
b) By how much does the volume of the container increase?
c) By what fraction does the water have to be compressed to prevent it from expanding more than the container?
d) What is the resulting pressure built up in the container? (Give your answer in pascals and atmospheres.)
(Assume the steel does not yield to the pressure of the water.)
[For water, $\beta \approx 3 \times 10^{-4}$ K^{-1} and $K \approx 2 \times 10^{10}$ Pa^{-1}.
For steel, $\beta \approx 3 \times 10^{-5}$ K^{-1}.]

Latent heats

5.17 The latent heat of vaporization of helium at a

pressure of 1 atm and a temperature of 4.2 K is 21.8 kJ kg^{-1}. The densities at 4.2 K of the liquid and vapour are 125 kg m^{-3} and 19 kg m^{-3} respectively. What proportion of the latent heat is involved in work against the interatomic attraction?

5.18 Use the data of the previous question to estimate the depth of the potential well (ε of figure 5.2) which results from the forces between two helium atoms. Give your answer in joules, electronvolts and kelvins.
[For helium, $A_r = 4$.]

5.19 A refrigerator uses ammonia as the *working fluid*. Estimate the mass circulation rate which is needed to extract heat at the rate of 20 W at −33 °C, the normal boiling point of the liquid. What is the corresponding volume circulation rate of the liquid?
[Latent heat of vaporization of ammonia = 1.4×10^6 J kg^{-1}.
Density of liquid ammonia at −33 °C = 685 kg m^{-3}.]

Surface tension

5.20 The surface tension of liquid ^{4}He at 1 K is 3.5×10^{-4} N m^{-1} and the liquid density is 145 kg m^{-3}. Estimate
a) the number of atoms per unit area in the surface,
b) the energy per bond in the liquid at this temperature.
(Give your answer to (b) in joules, electronvolts and equivalent kelvins.)

5.21 Calculate the capillary rise of water in a tube of internal diameter 0.1 mm.

It is proposed to construct a self-running pump to raise water through a height of 200 mm by taking such a fine bore tube and curving the top over so that the water, which has reached the top by capillary rise, drips out at the higher level. Comment on this proposal.
[Surface tension of water = 7.3×10^{-2} N m^{-1}.]

5.22 When a capillary tube of radius 0.5 mm is dipped into alcohol, it is found that the meniscus may be kept at the level of the outside liquid surface by applying a pressure of 9.0 mmH_2O to the top of the tube. What value for the surface tension of alcohol may be deduced from these observations?

How would you measure the pressure difference of 9.0 mmH_2O with adequate accuracy?

5.23 A thin walled glass tube has a radius of 0.5 mm and is open at both ends. The tube is filled with water and then held with its axis vertical. Some of the water runs out. What length remains?
[Surface tension of water = 7.3×10^{-2} N m^{-1}.]

5.24 A flat circular disc, 200 mm in diameter, is suspended from the centre of one face by a thread. If the disc is wetted by water, find the force required to pull the disc vertically upwards away from a water surface.

What would happen if water did not wet the material of the disc?
[Surface tension of water = 7.3×10^{-2} N m^{-1}.]

5.25 A mercury barometer has a tube of internal diameter 2 mm which is inverted over a relatively large dish of mercury. By how much does the height of the column differ from atmospheric pressure as a result of surface tension? Does the barometer read too high or too low?
[Surface tension of mercury = 0.47 N m^{-1}.
Density of mercury = 13.6 Mg m^{-3}.
Angle of contact of mercury to glass = 137°.]

5.26 It is possible to make porous glass: glass which contains a network of fine channels. Calculate the pressure which would be required to force mercury into such a glass if the channels have a diameter of 20 nm. (Give your answer in pascals and atmospheres.)
[For mercury, $\gamma = 0.47$ N m^{-1} and the angle of contact is about 137°.]

Vapour pressure: molecular origin, temperature dependence. Saturated and unsaturated vapours. Relative humidity. Thermionic emission.

5.27 The saturated vapour pressure of carbon tetrachloride varies with temperature as shown in the table.

t/°C:	0	10	20	30	40
p/mmHg:	33	56	91	140	215

a) Plot a graph of $\ln p$ against $1/T$ and deduce the latent heat of vaporization (i.e. fit the measurements to the approximate formula *5.12*.)
b) Plot a graph of $\ln (p/T)$ against $1/T$ and deduce the latent heat of vaporization (i.e. fit the measurements to formula *5.11*).
Both your graphs should be reasonably straight lines which shows how unimportant the pre-exponential temperature factor is in comparison with the rapidly varying exponential.
[You may wish to use the fact that $\ln x \approx 2.30 \lg x$.]

5.28 A long uniform tube, closed at one end, is inverted over mercury in a dish. The tube contains a little air and a small drop of benzene. The following observations are taken of H, the height of the top of the tube above the level of the mercury surface in the dish, and h, the height of the mercury column in the tube.

H/mm:	800	400
h/mm:	625	364

a) What determines the total pressure above the mercury in the tube?
b) How will this vary with the volume occupied by the gas?
c) Atmospheric pressure is 760 mmHg. What is the vapour pressure of the benzene?

5.29 On a day when the temperature is 20 °C the dew point is found to be 7.5 °C. Use the following data relating the saturated vapour pressure of water p_v to temperature t to derive an approximate value for the relative humidity.

t/°C:	0	5	10	15	20	25
p_v/mmHg:	4.58	6.54	9.21	12.78	17.51	23.69

5.30 Explain how thermionic emission of electrons from the surface of a hot metal is similar to evaporation from a liquid.

In measurements of thermionic emission from tungsten, the following values for the emitted current I at temperatures T were found:

T/K:	1520	1650	1790
I/μA:	0.23	2.7	32

Estimate the work function of tungsten.

Vapour pressure over curved surfaces. Nucleation in condensation and boiling; supercooling, superheating.

5.31 The saturated vapour pressure of water at 5 °C is 6.54 mmHg. What would be the radius of a water droplet which would be in equilibrium with water vapour at this temperature at a supersaturation of 10%?
[Surface tension of water at 5 °C = 7.5×10^{-2} N m^{-1}.
Relative molecular mass of water = 18.]

5.32 If condensation to form cloud normally occurs in the atmosphere at supersaturations of a few percent, what is the approximate effective radius of the condensation nuclei?

5.33 Explain the importance of nucleation in connection with boiling and condensation. Would you expect nucleation to be important in the change from liquid to solid?

The normal boiling point of nitrogen is 77.4 K, but when it is pure and is in a container with polished walls, it is found that boiling does not occur until the temperature reaches 80.5 K. The surface tension of liquid nitrogen is 8.5×10^{-3} N m^{-1}. Use the following data for the temperature variation of the saturated vapour pressure p_v to estimate the maximum size of nuclei in the liquid.

T/K:	77.36	79.24	84.27
p_v/mmHg:	760	939	1591

[1 atm = 760 mmHg = 1.01×10^5 Pa.]

5.34 Some water in a vessel of negligible thermal capacity is supercooled to −3 °C. The vessel is then thermally insulated and a small crystal of ice dropped in to nucleate freezing. When the system reaches equilibrium, what fraction of the water will have turned to ice?
[Specific heat capacity of water at 0 °C = 4.22 kJ K^{-1} kg^{-1}.
Latent heat of fusion of water = 333 kJ kg^{-1}.]

6 Thermal Radiation

6.1 GENERAL NATURE OF THERMAL RADIATION*

There are three mechanisms by which heat is transferred: conduction, convection and radiation. Conduction and convection both require matter to be present. In conduction, thermal motions diffuse through the substance (pages 60, 61), and convection involves bodily movement in a fluid. In contrast, radiation requires no intervening medium.

Radiation is the means by which we are warmed by the sun on a sunny day. The fact that the warmth stops at the same time as the sunshine if the sun is obscured by cloud, or an eclipse, shows that radiation travels in straight lines and that its speed is large. This suggests that thermal radiation might be electromagnetic radiation, similar to visible light. This is confirmed by spectroscopy: if a spectrometer is constructed out of components which do not absorb thermal radiation then energy is detected beyond the red end of the visible spectrum, that is, at longer wavelengths (figure 6.1). The spectrum does not consist of discrete spectral lines, as does the light from a mercury vapour lamp or hydrogen discharge tube, for example, (figure 6.2) but the energy is spread smoothly over a range of wavelengths. Both the intensity of radiation and the wavelength at which it is most intense vary with the temperature of the emitting body. (See figure 6.8, page 93.) With a source at room temperature, the radiation is most intense at wavelengths near 10 μm in the far *infra red*; while for the sun, with an effective surface temperature of about 6000 K, the maximum comes at about 480 nm in the visible part of the spectrum (blue) (figure 6.3).

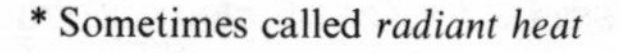

* Sometimes called *radiant heat*

6.2 DETECTORS FOR THERMAL RADIATION

Since thermal radiation carries energy, the simplest way of detecting it is to absorb it and measure the resulting temperature rise. The principle is illustrated by a *differential air thermometer* with one bulb blackened and one silvered (figure 6.4). Mercury is trapped

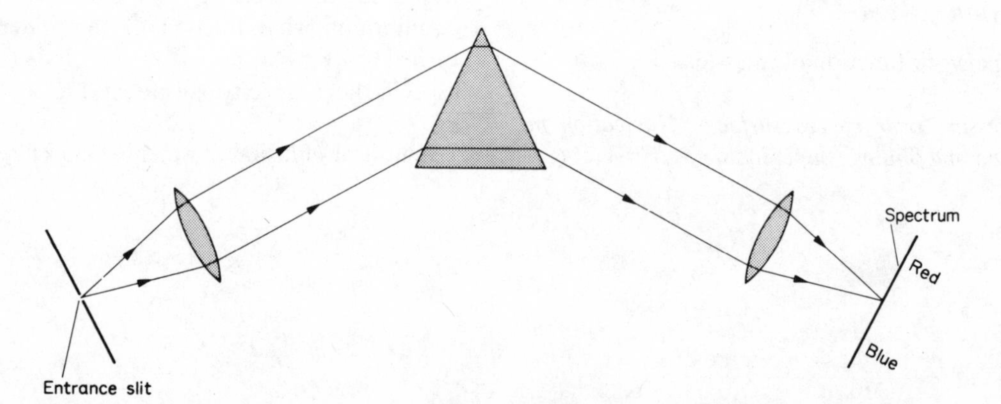

Figure 6.1 Thermal radiation from a hot body is detected beyond the red end of the visible spectrum. The spectrometer must be made of components which do not absorb the long wave thermal radiation. Glass is opaque in the infra-red; but quartz is not. A prism spectrometer with quartz components is therefore suitable for examining thermal radiation. Alternatively, a spectrometer with a reflection grating and surface silvered mirrors may be used.

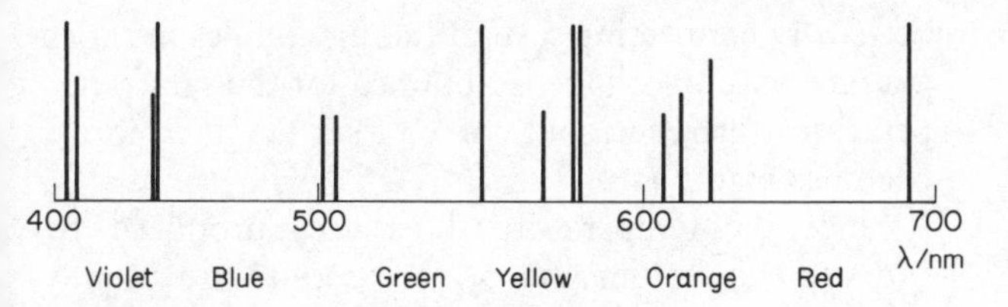

Figure 6.2 The line spectrum of a mercury vapour discharge lamp in the visible part of the spectrum. The heights of the lines give a rough indication of the intensities of the spectral components.

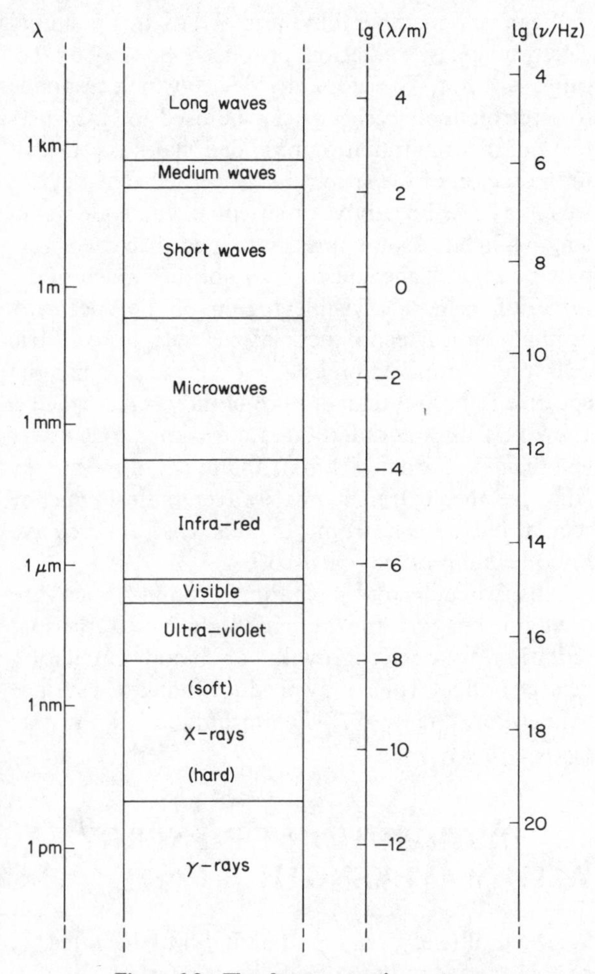

Figure 6.3 The electromagnetic spectrum

in a capillary between two bulbs of equal volume containing air. Changes of ambient temperature (the temperature of the surroundings) cause equal changes of pressure in the bulbs and the mercury does not move. But the black surface absorbs radiation while the silvered one reflects it, so, if radiation is incident, the blackened bulb warms more than the silvered one and the mercury moves.

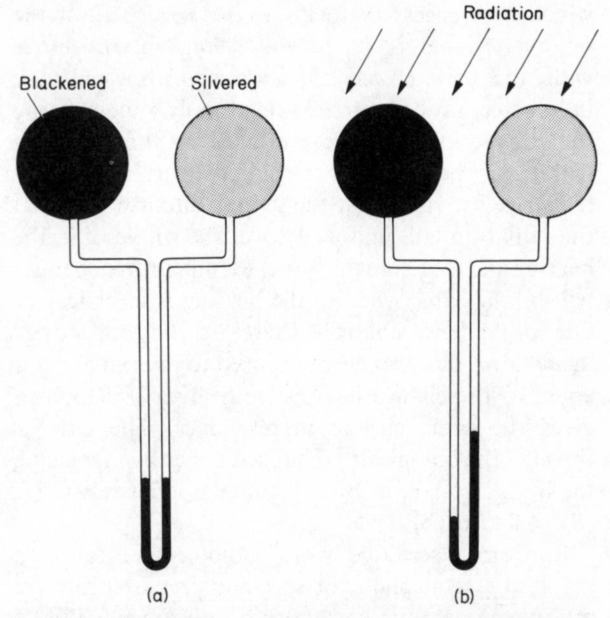

Figure 6.4 A differential air thermometer. Changes in ambient temperature affect the pressure in the bulbs equally and the mercury does not move (a). When radiation is incident, the blackened bulb is heated more than the silvered one and the mercury moves (b).

Such an arrangement is crude, clumsy and insensitive. For serious work, *bolometers* are usually used. These are devices in which the temperature rise is measured by finding the change in electrical resistance of a conductor. A thin blackened strip of aluminium foil makes a good, simple bolometer. The resistance can be measured by connecting the foil in a Wheatstone bridge circuit (figure 6.5). For experiments in

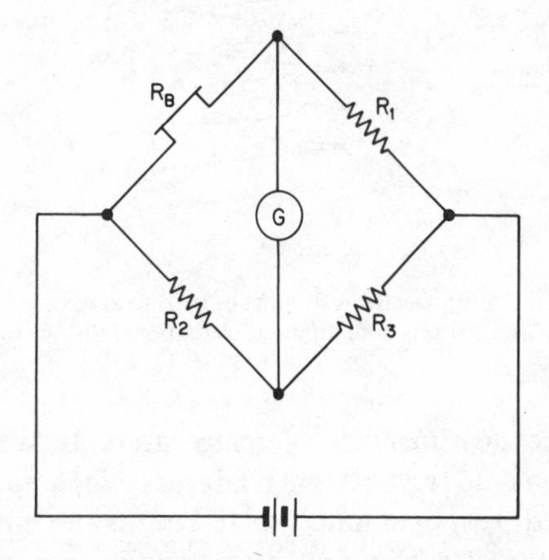

Figure 6.5 A bolometer connected in a Wheatstone bridge circuit. The bridge balances when $R_B/R_1 = R_2/R_3$. For small deviations from balance, the current through the galvanometer is proportional to the change in resistance of the bolometer.

which it is necessary to know the *sensitivity* of the bolometer, that is, its rate of change of resistance with incident power (Δ(resistance)/Δ(power)), the bolometer may be calibrated very simply by comparing the change in resistance produced by the radiation with that produced by increasing the current through the strip. Provided that the strip is uniform and that the radiation falls uniformly over the whole of it, the heating and therefore the way the temperature changes will be the same whether the heating is electrical or due to incident radiation. Therefore, the total power incident on the strip when exposed to the radiation is equal to the change in resistive heating (I^2R) which gives the same change in resistance. The current through the bolometer is changed either by increasing the bridge voltage or by reducing the series resistance (R_1 of figure 6.5).

For more sensitive work, bolometers are often made of a semiconductor specially prepared to have a large temperature coefficient of resistance (dR/dT). When cooled to low temperatures, they can be extremely sensitive. *Semiconducting bolometers* with a receiving area of 1 mm^2 operating at a few kelvins can detect signals of less than 10^{-14} W, roughly equivalent to the radiation from a candle 20 miles away!

Another useful radiation detector is the *thermopile* which consists of a number of thermocouples connected in series and arranged so that one set of junctions is exposed to the radiation while the other is not (figure 6.6). Changes of ambient temperature

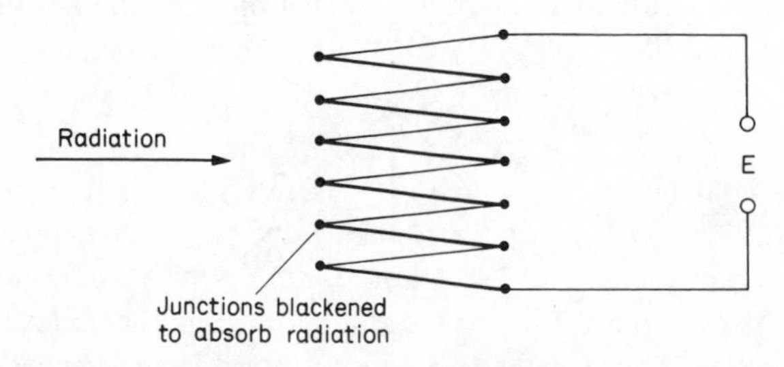

Figure 6.6 A thermopile. A number of thermocouples is connected in series and arranged so that the radiation falls on one set of junctions.

produce no *difference* of temperature between the junctions, so voltage only appears when radiation causes the exposed junctions to become warmer than the others. Usually, the thermocouples are made of conductors which give a large thermoelectric e.m.f. Traditionally, antimony and bismuth are used, but nowadays specially prepared semiconductors are better. By connecting several thermocouples in series a larger output voltage is obtained for the same temperature difference, but, of course, the thermopile becomes bigger too.

Since the temperature differences produced by radiation in bolometers or thermopiles are usually small, Newton's law of cooling applies to the heat loss to the surroundings (page 26), and the change in resistance or the output voltage in the case of thermopiles is linearly proportional to the incident power.

We must remember that thermal radiation is simply electromagnetic radiation produced by a body because it is hot. Therefore any device which responds to electromagnetic energy can be used for measurement of thermal radiation provided that it is sensitive in the region of the spectrum concerned. The devices we have described above are useful over a wide wavelength range, but sometimes more specialized detectors may be used. If the source is so hot that much of the radiation is in the visible region of the spectrum, ordinary optical techniques may be used: photoelectric cells, photographic film. At the opposite extreme, the effective radiation temperature of interstellar space is a few kelvins; the radiation is most intense at wavelengths of the order of 1 mm, and much of the energy falls in the extreme microwave radio-frequency region of the spectrum. In this case, microwave radio techniques may be used.

Sensitive detectors operating around 10 μm are used for imaging the human body by its thermal radiation to detect growths or blood circulation defects, both of which may produce changes in surface temperature (figure 6.7). The technique is known as *thermography*.

6.3 DISTRIBUTION OF ENERGY WITH WAVELENGTH

We have already remarked that both the intensity of the energy radiated and the way in which it is distributed with wavelength depend on the temperature of the emitting surface. To describe precisely how energy is distributed with wavelength, we define a quantity called *spectral emissive power* e_λ:

$e_\lambda\, d\lambda$ is the energy emitted per second per unit area of surface in radiation with wavelengths between λ and $\lambda + d\lambda$.

The units of e_λ are $J\,s^{-1}\,m^{-2}\,m^{-1} = W\,m^{-3}$. The

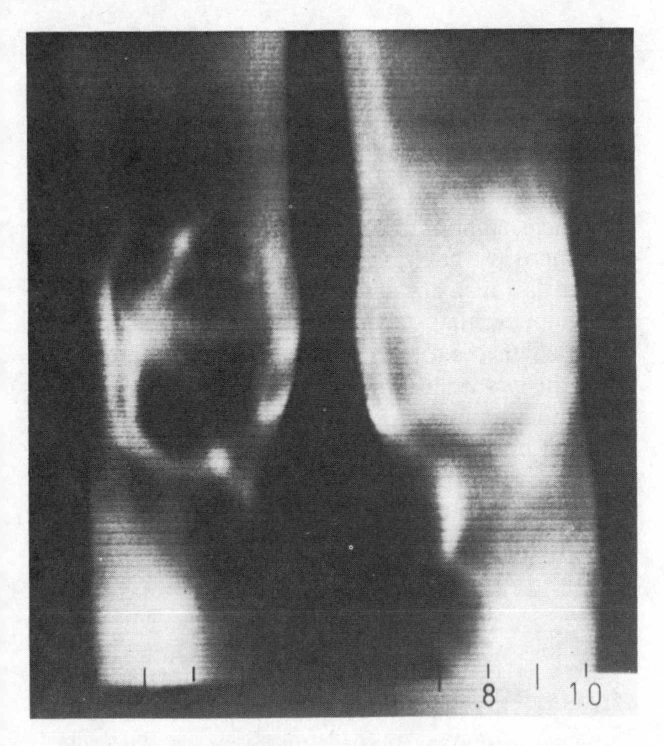

Figure 6.7 A thermogram of the knees of a woman with chronic rheumatoid arthritis. The picture shows the front view of the knees. The cooler areas appear dark and the warmer areas light. The left knee is affected, the inflammation causing the skin temperature to rise above that of the other knee by about 5 °C. (Thermogram by courtesy of AGA Infrared Systems AB.)

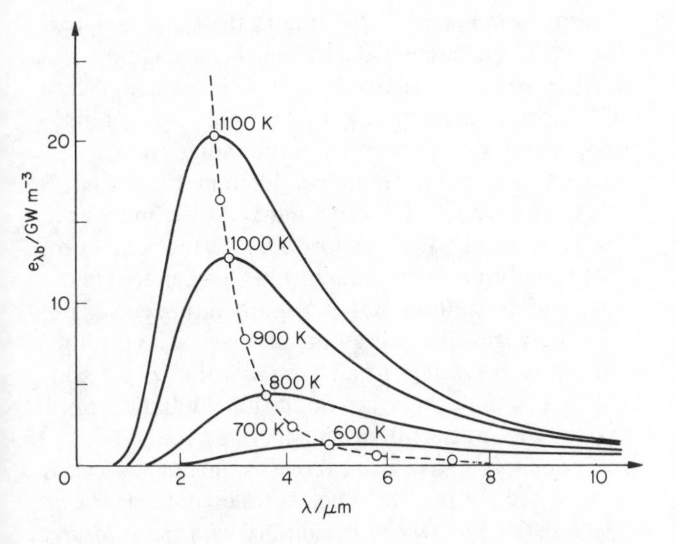

Figure 6.8 The spectral emissive power of a black surface. Curves are shown for several surface temperatures. The broken line shows the variation with wavelength and temperature of the peaks in the curves.

last m^{-1} corresponds to 'per unit wavelength interval' and is cancelled dimensionally by the dλ in the definition. e_λ is therefore a distribution function exactly like that which we encountered earlier for the speeds of molecules in a gas (page 53). Curves of e_λ for a black surface (see page 95) are shown in figure 6.8.

What is the mechanism by which thermal radiation is produced? Radiation is always the result of movement of charge. Just as the currents flowing in a radio aerial connected to a transmitter result in radiation of radio waves, so also does motion of charge in atoms or molecules result in radiation. With thermal radiation, this motion of charge is caused by thermal agitation in the material. As the temperature is increased, the agitation becomes more energetic so that we would expect:

a) more total power to be emitted, and
b) that the radiation would shift to shorter wavelengths. This is a result of quantization (page 51): the energy of a photon (a quantum of electromagnetic radiation) is $h\nu = hc/\lambda$, where h is the Planck constant and ν and λ are the frequency and wavelength of the radiation. As more energy is available from the thermal motions to excite the production of photons, we expect not only *more* photons to be produced but also *more energetic* ones. So the radiation will shift to higher frequencies (shorter wavelengths).

We see in figure 6.8 that these ideas are in agreement with the observed spectra.

6.4 PRÉVOST'S THEORY OF EXCHANGES

What we have already said about the way thermal radiation depends on temperature suggests that all bodies must emit thermal radiation all the time. If the radiation is caused by thermal motions in the material, then there will always be radiation as long as there is thermal motion, that is, except at absolute zero, 0 K, where thermal motion would cease. This means that when a body is in thermal equilibrium with its surroundings, the equilibrium is *dynamic*: the body continually radiates to the surroundings and continually absorbs radiation from the surroundings. This is *Prévost's theory of exchanges* which he first proposed (1792) after observing that objects initially in thermal equilibrium with their surroundings became cooler if a cold object was placed nearby.

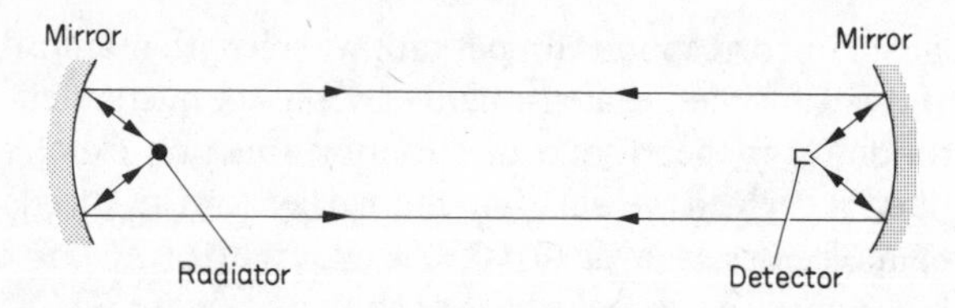

Figure 6.9 An arrangement for demonstrating Prévost's theory of exchanges

Figure 6.9 shows a simple arrangement for demonstrating Prévost's theory of exchanges. It also illustrates reflexion and rectilinear propagation of thermal radiation. Two concave mirrors, some distance apart, face one another. At the focus of one is an object to act as radiator (a blackened metal sphere, for example). A detector is at the focus of the other. (The optics may be set up using a light source. Since both light and thermal radiation consist of electromagnetic waves, if the adjustment is right for one it will be right for the other also.) Radiation from the sphere is reflected by the left mirror into a parallel beam which is collected by the other mirror and focused on the detector. Radiation from the detector follows the reverse path and is focused on the sphere. When the whole system is in thermal equilibrium, there is no *net* energy exchange between radiator and detector and the output from the detector is zero. If the sphere is now warmed, the detector gives an output corresponding to a net absorption of radiation. If the sphere is cooled (by dipping in ice-cold water or liquid nitrogen) then the meter is deflected to the other side of zero showing that the detector is cooling because it is receiving less radiation from its surroundings than before. To work well, the mirrors need to be surface-silvered or of polished metal, because glass absorbs radiation in the far infra-red which is the region of the spectrum involved for room temperature radiation.

6.5 EQUILIBRIUM RADIATION

Suppose we have a closed (opaque) container whose walls are at a uniform temperature. Such a container is called an *equal temperature enclosure.* The space inside the enclosure will be filled with radiation going to and fro as a result of continual emission, reflexion and absorption by the inner surfaces. After a short time, we would expect the radiation to come into thermal equilibrium with the walls of the enclosure: there would be no net flow in any direction, no net absorption by any part of the walls, and the energy present in the radiation would be constant with time. The radiation is then called *equilibrium radiation.* We shall now prove that the nature of the radiation (how the energy in it varies with wavelength and temperature) is independent of the nature of the walls of the enclosure.

Suppose we have two equal temperature enclosures, A and B, which are initially at the same temperature, and suppose that the energy in the radiation around wavelength λ is greater in A than in B. We now connect the two enclosures by a tube in which there is a filter to let through a narrow band of wavelengths centred on λ (figure 6.10). Then, since we have supposed that the radiation around this wavelength is more intense in A than in B, more energy will pass through the filter from A to B than in the reverse direction. But if this happens, A will cool down and B will warm up. But we know that if two objects at the same temperature are put in thermal contact (regardless of the means of thermal contact) then they do not diverge in temperature.* Being at the same temperature means precisely that there will be no change when thermal contact is established—that is the significance of temperature (section 1.3). Therefore, our original supposition that the energy in the radiation at a particular wavelength could be different in A and B when they were at the same temperature, must have been impossible. We therefore come to the conclusion that equilibrium radiation must be independent of the nature of the walls of the equal temperature enclosure. We know that the energy in thermal radiation does vary with wavelength and temperature, so we conclude that the energy in equilibrium radiation must depend on wavelength and temperature only.

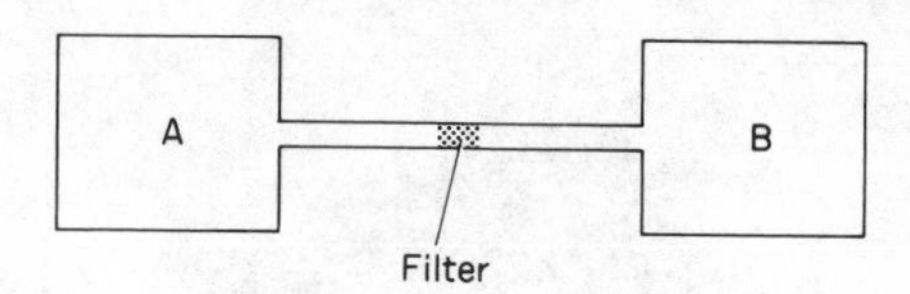

Figure 6.10 Proof that the nature of equilibrium radiation does not depend on the nature of surfaces

The proof we have given above would not be affected if the equal temperature enclosures contained objects *provided that they were in equilibrium with the enclosure.*

* It is just this sort of thing that the second law of thermodynamics says does not happen. See section 7.1.

The energy in the radiation would have to be just the same, and, in particular, there could be no net flow of radiant energy in any direction inside the enclosures: the radiation would be *isotropic*. This means that it would be impossible to distinguish objects at equilibrium in an equal temperature enclosure by measurements on the radiation. Everywhere you looked, there would be uniform brightness and colour. The only reason you can see the print on this page is that this room is not an equal temperature enclosure. In particular, there are sources of high-temperature radiation (producing visible light) and the print is distinguished by its different reflecting properties as compared with the paper.

We will now formally define a *black body* as one which absorbs all radiation of all wavelengths which falls on it. Any body in equilibrium in an equal temperature enclosure must radiate as much energy at each wavelength as it absorbs (to preserve the nature of the radiation) and this must be true for black bodies also. But a black body absorbs *all* incident energy, so the radiation *emitted* by the black body must be *identical* with the (equilibrium) radiation incident on it. What do we mean by identical? We mean that the energy must be distributed with wavelength and depend on temperature in exactly the same way. For this reason, equilibrium radiation is also called *black body radiation.*

Also for this reason, most experiments on equilibrium radiation are done by making measurements on the radiation emitted by a black body. A blackened surface, such as one covered with soot from a smoky flame, can be a good approximation to a genuine black surface. It is better, though, to construct an equal temperature enclosure with a small hole in it through which radiation may escape. If one imagines radiation *entering* through the hole from outside (figure 6.11), then, even though there may be some reflection from the inner surfaces, the proportion of the energy entering which will escape back through the small hole will be negligible. Thus, looking into the hole, the cavity will be very nearly perfectly black; which means that radiation escaping through the hole will be nearly perfectly black also.

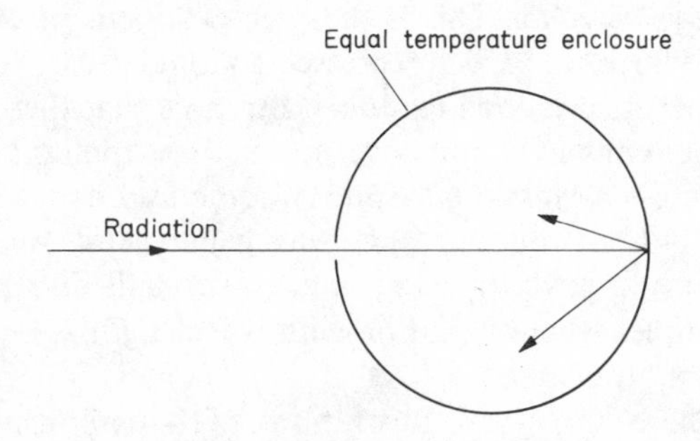

Figure 6.11 An equal temperature enclosure as a black body. If radiation enters an equal temperature enclosure through a small hole, a negligible proportion will escape again even if the walls are not black. The interior therefore appears black to radiation entering. Therefore radiation escaping through the hole will be black.

6.6 INFLUENCE OF THE NATURE OF A SURFACE

In the last section we discussed some of the properties of ideal black surfaces. Few real surfaces approach ideal blackness. Most reflect a considerable proportion of incident radiation, many are 'coloured' in that the proportion they reflect is different for different wavelengths. We have shown that when radiation is in equilibrium in an equal temperature enclosure, its character is independent of the nature of the walls. Then the walls must have special properties to ensure this. In particular, if they absorb strongly at some wavelength, they must also radiate strongly at that wavelength, or else the radiation leaving them would be different from that arriving.

We have already mentioned the differential air thermometer which depends on different absorbing properties of surfaces for its operation (page 91). Different radiating properties are usually illustrated with *Leslie's cube*, a cubic metal box which can be filled with hot water and whose sides are differently treated so that their emissive properties may be compared. The cube is usually set up on a turntable so that any side can be rotated to face a detector which is commonly a thermopile (figure 6.12). Traditionally, one face is blackened, one polished, one covered with white paper and one left untreated. It is found that the black surface emits most strongly and the polished one, which is the best reflector, least strongly. It therefore seems likely that *good absorbers are good emitters.* This idea is expressed more precisely in Kirchhoff's law.

Consider a body in equilibrium inside an equal temperature enclosure. We know that the energy *incident* on it is identical to the energy which would be *emitted* by a black surface at the same temperature. So the incident energy per area per second in

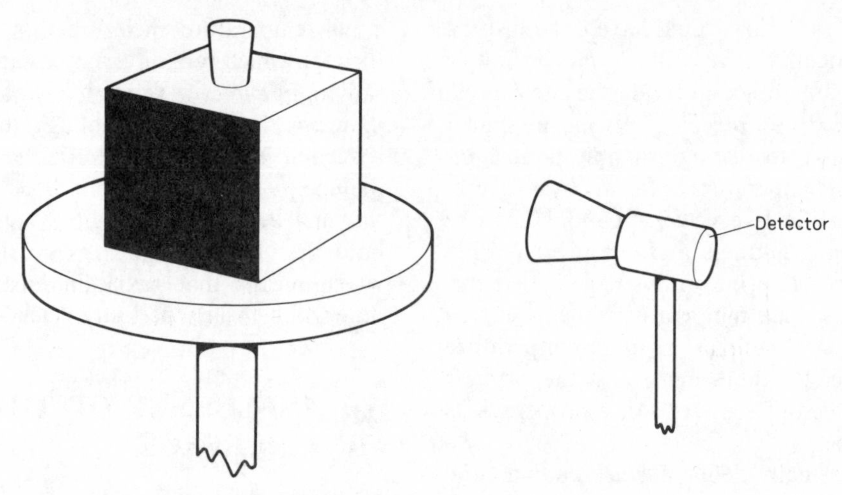

Figure 6.12 Leslie's cube

wavelengths between λ and $\lambda + d\lambda$ is $e_{\lambda b}\, d\lambda$ where $e_{\lambda b}$ is the spectral emissive power of a black body. We define *spectral absorptivity* α_λ as the fraction of the energy incident at wavelength λ which is absorbed.* (Here, the significance of the word 'spectral' is to indicate that we are referring to a particular wavelength.) Then the energy absorbed in wavelengths between λ and $\lambda + d\lambda$ is

$$\alpha_\lambda e_{\lambda b}\, d\lambda$$

The same amount must be radiated if the nature of the radiation is to be unchanged. Hence, the spectral emissive power of the surface e_λ must be given by

$$e_\lambda\, d\lambda = \alpha_\lambda e_{\lambda b}\, d\lambda$$

and cancelling the $d\lambda$'s,

$$e_\lambda = \alpha_\lambda e_{\lambda b} \qquad (6.1)$$

But $e_{\lambda b}$ depends on temperature and wavelength only (section 6.5). Hence the ratio e_λ/α_λ depends on wavelength and temperature only. In particular, it does not depend on the nature of the surface. This result is *Kirchhoff's law*:

The ratio of the spectral emissive power to the spectral absorptivity is a universal function of wavelength and temperature only.

(This means that the ratio depends on wavelength and temperature only, and that the dependence is the same whatever the surface and whatever the conditions.)

Since a black body is the best possible absorber (it has $\alpha_\lambda = 1$ for all λ), Kirchhoff's law shows that it must also be the best possible emitter. If we define *spectral emissivity* ε_λ as the ratio of the spectral emissive power of the surface concerned to that of a black surface:

$$\varepsilon_\lambda = \frac{\text{spectral emissive power of surface concerned}}{\text{spectral emissive power of a black surface}}$$

then we have, substituting in *6.1*,

$$e_\lambda = e_{\lambda b}\varepsilon_\lambda = e_{\lambda b}\alpha_\lambda$$

so

$$\varepsilon_\lambda = \alpha_\lambda \qquad (6.2)$$

Thus, *spectral emissivity and spectral absorptivity must always be equal.* This is the precise statement of the fact which we earlier expressed colloquially by 'good absorbers are good emitters'. But note that the identity of emissivity and absorptivity only applies *to the same wavelength* (figure 6.13). A coloured surface has α_λ and ε_λ varying with wavelength and we will generally have $\alpha_{\lambda_1} \neq \alpha_{\lambda_2} = \varepsilon_{\lambda_2}$. We shall give some examples where variation with wavelength is important a little later.

There are many illustrations of Kirchhoff's law. The radiation temperature of interstellar space is a few kelvins. On a clear night, the earth's surface radiates into space but receives very little radiation back from space. If there is no wind to stir the air and keep the surface warm, the surface temperature

* Being a ratio of energies, the absorptivity is dimensionless. Soon we shall define emissivity which is also dimensionless. In both cases we use a Greek letter in contrast to our use of a Roman letter *e* for the dimensional quantity emissive power.

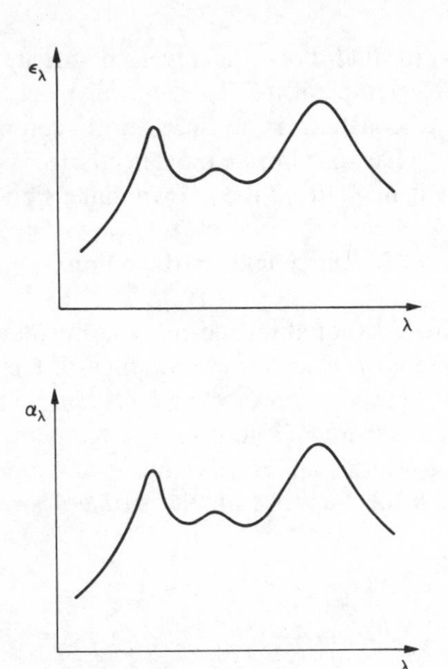

Figure 6.13 Identity of spectral emissivity and spectral absorptivity. Spectral emissivity and spectral absorptivity may vary with wavelength, but at each wavelength, $\varepsilon_\lambda = \alpha_\lambda$.

drops and dew or frost is formed. However, water is a good absorber in the infra-red (where the earth's radiation mainly lies) so that any cloud not only absorbs the radiation from the earth's surface, but also, since the cloud is not much colder than the surface, radiates a consisderable amount of energy back. The net energy loss is much smaller, and it is only exceptionally that frost or dew forms.

We will consider two examples where the change of α_λ with wavelength is important:

In the Leslie's cube demonstration of the emitting properties of surfaces, the black surface (a good absorber) is the best emitter, and the polished surface (a good reflector) the poorest emitter. But, it turns out that the white paper surface is nearly as good an emitter as the black surface! The paper is white because it reflects *visible* radiation well (and emits it badly); but between the visible part of the spectrum and the infra-red where the thermal radiation mainly lies, the paper becomes a good absorber (and emitter): in the infra-red, white paper is nearly black!

Greenhouses provide the second example. Greenhouses can keep the plants inside warmer than the surroundings. How do they do this? Glass is transparent in the visible but opaque (a good absorber) in the infra-red. The sun is a high temperature source (corresponding to about 6000 K), so much of the energy it radiates is in and near the visible part of the spectrum. This is transmitted by the glass and largely absorbed by the objects inside the greenhouse. The objects inside radiate, but they are much cooler than the sun (!), say about 300 K, so their thermal radiation is in the far infra-red and is intercepted by the glass. The glass thus lets in the short wavelength radiation but acts as a 'radiation shield' to outgoing long wavelength radiation. The one-way effect is the result of the wavelength change when the energy is absorbed and re-emitted from a lower temperature source (figure 6.14).

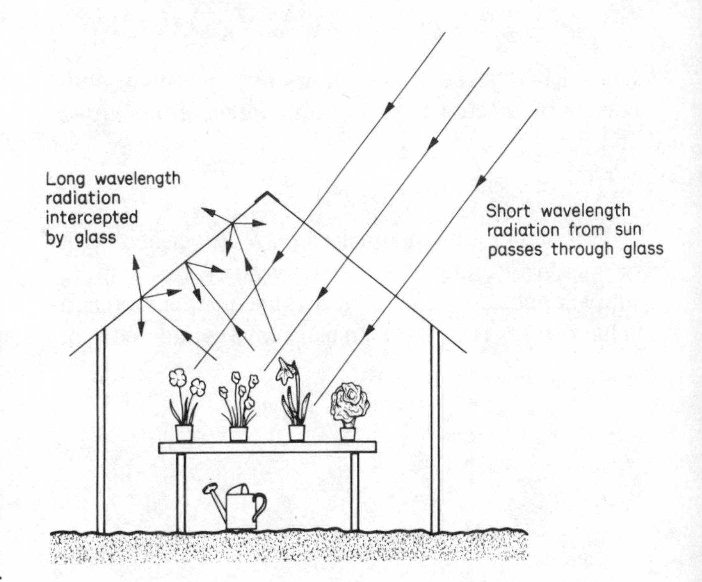

Figure 6.14 A greenhouse

6.7 THE STEFAN–BOLTZMANN LAW

We now turn to discuss some other aspects of the energy emitted by a hot body.

Figure 6.8 shows how the energy is distributed with wavelength. The *total* energy emitted is the area under the curves. We call this the (total) *emissive power* of the surface, e:

$$e = \int_0^\infty e_\lambda \, d\lambda \qquad (6.3)$$

The units of emissive power are $J\,s^{-1}\,m^{-2} = W\,m^{-2}$.

The emissive power of a black surface is found experimentally to be proportional to the fourth power of its thermodynamic temperature:

$$e_b = \sigma T^4 \tag{6.4}$$

This is *the Stefan–Boltzmann law.** Remember that e_b is energy per area per second, so the law could also be written

$$\frac{dQ}{dt} = \sigma A T^4$$

where Q is the total thermal energy (heat) radiated and A is area of surface.

The constant of proportionality σ is the *Stefan–Boltzmann constant.* Its value is

$$\sigma = 5.67 \times 10^{-8}\ \mathrm{W\,m^{-2}\,K^{-4}}$$

If a surface is not black, it is often possible, with reasonable accuracy, to use an average emissivity ε and put

$$e = \varepsilon e_b = \varepsilon\sigma T^4 \tag{6.5}$$

The Stefan–Boltzmann law may be verified with reasonable accuracy in the laboratory using quite simple apparatus. A suitable arrangement is sketched in figure 6.15. The source is a blackened cylinder, which

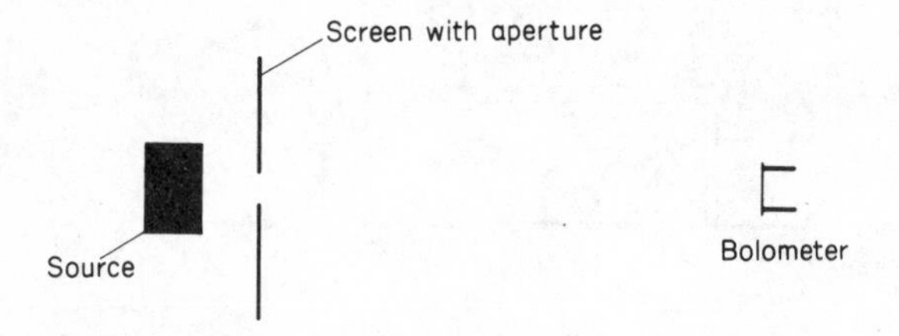

Figure 6.15 The essential elements of an arrangement for verification of the Stefan–Boltzmann law

may be either solid, with a hole drilled to take a thermometer, or a can filled with high boiling point oil. The source is heated electrically or with a gas burner. The detector is a bolometer. To define the area of emitting surface seen by the bolometer, a screen is placed in front of the source with a hole through which part of the source is exposed. With the source at room temperature, the whole system is in thermal equilibrium. When the source is heated, the rise in temperature of the bolometer corresponds to the *increase* in incident power. Verification of the T^4 dependence is straightforward because it is only necessary to find how the received energy varies with source temperature. To determine the Stefan–Boltzmann constant is a little more complicated because one has to calibrate the detector (see page 92) and also calculate (from the known dimensions of the apparatus) what fraction of the power from the exposed area of source falls on the bolometer.

* Also called *Stefan's Law*

6.7.1 There is a common and culpable mistake which is often made in calculating exchange of energy by radiation between a body and its surroundings. Consider a sphere of radius r_1 at temperature T_1 enclosed in another sphere of radius r_2 at temperature T_2 (figure 6.16). Suppose all the surfaces are black.

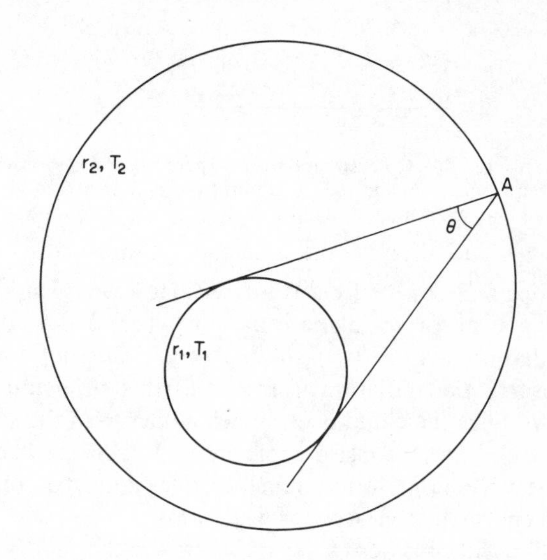

Figure 6.16 Energy exchange by radiation between a sphere and its surroundings

The erroneous argument goes as follows:

i) The power emitted outwards by the inner sphere is $4\pi r_1^2\sigma T_1^4$.
ii) The power emitted inwards by the outer sphere is $4\pi r_2^2\sigma T_2^4$.
iii) Therefore, the net rate of exchange of energy is

$$4\pi\sigma(r_1^2T_1^4 - r_2^2T_2^4)$$

If one stops to check whether this result is reasonable (one should *always* do this at the end of any calculation), it is obvious nonsense: if we put $T_1 = T_2$ there is a net flow of energy. In the first place, we know that equality of temperature implies thermal equilibrium and therefore no heat flow. Also, if the result were true, we could use the flow of heat to power an engine (say, produce a temperature difference and then extract power with a thermocouple) to produce perpetual motion.

The point is, that, although all the power radiated by the inner sphere strikes the outer sphere and is absorbed by it, only a proportion of the energy radiated by the outer sphere is absorbed by the inner; the rest misses the inner sphere and returns to another part of the outer. In figure 6.16 it is only the radiation within the angle θ from the point A which is intercepted by the inner sphere. Of course, one could go through the geometry to calculate the proportion which reaches the inner sphere. (It is difficult enough if the spheres are concentric, and becomes very difficult if they are not because the proportion varies, depending on where the energy is radiated from.) But the sensible thing is to use the fact that there cannot be any exchange when the temperatures are the same. We also know that all the power calculated in step (i) above does reach the outer sphere. Then to make the net exchange zero when the temperatures are the same, the area factor for the radiation *to* the inner sphere must also be $4\pi r_1^2$. Thus, the net exchange rate is

$$4\pi r_1^2\sigma(T_1^4 - T_2^4) \qquad (6.6)$$

The same argument applies whatever the shape of the outer container. Equation *6.6* therefore applies whatever the surroundings, provided, of course, that they are black and all at a uniform temperature T_2.

6.7.2 For small differences of temperature between a body and its surroundings, the exchange of radiant energy obeys Newton's law (page 26). That is, the net rate of energy flow is proportional to the temperature difference. This follows by factorizing the difference of the fourth powers of the temperatures:

$$\begin{aligned} T_1^4 - T_2^4 &= (T_1^2 - T_2^2)(T_1^2 + T_2^2) \\ &= (T_1 - T_2)(T_1 + T_2)(T_1^2 + T_2^2) \\ &\approx (T_1 - T_2)4T^3 \end{aligned}$$

where we have approximated by putting $T_1 = T_2 = T$, the mean temperature, except in the first bracket where we have the difference of the two temperatures.

6.7.3 Thermal radiation carries energy and takes time to travel, so there will always be energy in transit. This is why the presence of thermal radiation always results in there being a certain density of energy ($J\ m^{-3}$) in any space. We may use Stefan's law to estimate a typical value for the density of energy in the form of thermal radiation. Consider a black surface at temperature T. Each square metre of surface emits σT^4 joules per second. This energy will be radiated over all directions; but let us simplify the calculation by supposing that it is all emitted normally to the surface. At the end of one second, the energy emitted at the start of that second will have travelled c metres. The energy emitted per second, namely σT^4, will therefore be spread out over a volume $c \times (1\ m^2)$ (figure 6.17). Then the energy per cubic metre is

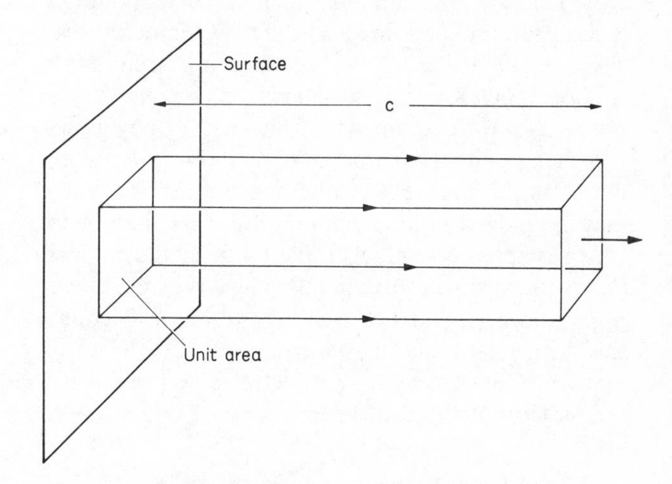

Figure 6.17 Approximate calculation of the energy density in equilibrium radiation

$\sigma T^4/c$. Evaluating this for $T = 290$ K, we obtain an energy density of $1.3 \times 10^{-6}\ J\ m^{-3}$. This energy density is small because the speed of light is large so that the energy is thinly spread out. (During the second, the total energy emitted is 400 J.) The exact calculation of the energy density inside an equal temperature enclosure (which takes into account the range of angles of emission and the presence of radiation from other parts of the enclosure) gives an energy density four times larger than is obtained by the simplified calculation above.

6.8 WIEN'S DISPLACEMENT LAW

We have discussed how the *total* power emitted depends on temperature. We now turn our attention to the way in which the wavelength of maximum intensity, the peak in the curves of figure 6.8, shifts with temperature. We have already given a qualitative explanation of this shift (page 93) in terms of the energy of thermal excitation. Experimentally, it is found that the wavelength of maximum intensity λ_{max} is inversely proportional to temperature. That is, $\lambda_{max}T$ = constant. This is *Wien's displacement law*. The value found experimentally (and also given by the fundamental theory of thermal radiation, which we shall refer to in the next section) is

$$\lambda_{max}T = 2.90 \text{ mm K} \qquad (6.7)$$

This is the relation we have used when giving the region of the spectrum in which thermal radiation is most intense for sources at different temperatures: room temperature, $\lambda_{max} \approx 10\ \mu m$, far infra-red; sun's surface (6000 K), $\lambda_{max} \approx 480$ nm, blue; etc. It is obvious that the colour with which a hot body glows will change with the temperature of the body. Between about 500 °C and 1500 °C the temperature of a body may be judged quite accurately by the colour with which it appears to glow. Table 6.1 gives the colours for various temperatures in this range. We note that a body appears white-hot well before the peak enters the visible part of the spectrum. At 1500 °C ($\lambda_{max} = 1.6\ \mu m$) there is already sufficient blue present for the radiation to appear white.

Table 6.1 How the apparent colour of a glowing body depends on temperature

t/°C	*Colour*
500	red, just visible in daylight
700	dark red
900	bright red ('cherry'-red)
1100	orange
1300	yellowish-white
1500	dazzling white

The eye has a remarkable ability to compensate for changes in the colour quality of light. We generally judge the colours of objects in much the same way whether we view them by daylight or by artificial light, yet the light from ordinary tungsten filament bulbs is very yellow in comparison with daylight. Colour films do not have this ability to compensate and that is why different films are available for different lighting conditions. The manufacturers design a colour film to give a pleasing colour balance when used with lighting of a certain 'colour temperature'. Normal daylight film is adjusted for a colour temperature of 5000 K, while artificial light films (for use with tungsten filament photofloods) are adjusted for about 3400 K. Electronic flash tubes are filled with xenon which gives a colour balance very similar to daylight, and so may be used with daylight film without compensating filters.

6.9 THE BIRTH OF QUANTUM THEORY

Towards the end of the last century various attempts were made to derive a formula to describe how the energy radiated by a black body depends on wavelength and temperature, that is, to derive a formula for $e_{\lambda b}$. These attempts, which were based on classical physics, failed. Classical ideas did give information about the *form* of the equation; but no argument based on them was able to lead to a satisfactory explicit formula.

At the turn of the century, Planck discovered an equation which fitted the experimental results; but the equation could only be derived if it was assumed that *the energy of oscillators is quantized*: the energy cannot take any value but must change in steps of a certain size. At each energy step a *quantum* of energy is absorbed. The size of the quantum is $h\nu$ where ν is the frequency of the oscillator and h a constant—the *Planck constant* (see also section 3.5.7). This was a revolutionary idea and must, at first, have seemed utterly ridiculous to Planck's contemporaries. Nothing in their everyday experience could suggest that quantization existed. h is so small (6.63×10^{-34} J s^{-1}) that the quantum of energy of any macroscopic oscillatory system (like a pendulum) is always minute compared with any measurable change in the total energy. As a result, it will always *appear* that the energy of macroscopic oscillators can take any value: the quantization is so fine-grained that it is unobservable. Planck's paper appeared in 1901. Four years later, Einstein published a paper showing that the same assumption could explain the photoelectric effect (the ejection of electrons from metals by light). These two historic papers* established quantum theory, which is a cornerstone of modern physics.

* M. Planck, *Ann. d. Physik*, **4**, 553 (1901)
A. Einstein, *Ann. d. Physik*, **17**, 132 (1905).

The *Planck radiation law* gives the energy density in black body radiation. The spectral emissive power of a black body follows immediately from it. The result is

$$e_{\lambda b} = \frac{2\pi hc^2\lambda^{-5}}{\exp(hc/\lambda kT) - 1} \qquad (6.8)$$

This is the formula which gives the curves of figure 6.8.

The Stefan–Boltzmann and Wien displacement laws follow immediately from *6.8*. Integrating $e_{\lambda b}$ over all wavelengths gives the Stefan–Boltzmann law for the (total) emissive power of a black body:

$$e_b = \left(\frac{2\pi^5k^4}{15c^2h^3}\right)T^4$$

The factor in brackets is the Stefan–Boltzmann constant:

$$\sigma = \left(\frac{2\pi^5k^4}{15c^2h^3}\right) = 5.67 \times 10^{-8}\ \mathrm{W\,m^{-2}\,K^{-4}}$$

Differentiating *6.8* with respect to λ and putting the differential equal to zero gives an equation for λT at the maximum in the black body radiation curves. The solution of the equation gives equation *6.7* which is the Wien displacement law.

PROBLEMS

General nature of thermal radiation

6.1 What experiments would you do to demonstrate the electromagnetic nature of thermal radiation?

Detectors: bolometers, thermopile, optical and radio techniques

6.2 When a hand is held a little way in front of a bolometer containing 30 antimony–bismuth thermocouples in series, a thermoelectric voltage of 75 μV is developed. How much are the exposed junctions warmed by the radiation from the hand?
[Sensitivity of antimony–bismuth thermocouple $\approx 4.7\ \mathrm{mV\,K^{-1}}$.]

6.3 A bolometer is connected in a Wheatstone bridge circuit as shown in figure 6.5. The cells have an e.m.f. of 4 V and negligible internal resistance. With no radiation falling on the bolometer, the bridge balances with the resistances set as in *A* or *B* below. With radiation incident on the bolometer, the bridge balances with the settings shown in *C*. What is the total power falling on the bolometer in *C* assuming that all the incident radiation is absorbed?

	R_1/Ω	R_2/Ω	R_3/Ω
A	1000	201.7	20 000
B	500	203.1	10 000
C	1000	203.3	20 000

What accuracy could you claim for the result with measurements of this accuracy?

Distribution of energy with wavelength; spectral emissive power

6.4 A car headlamp bulb is immersed in water contained in a glass beaker. When the lamp is switched on the temperature is found to rise at $0.0173\ \mathrm{K\,s^{-1}}$. When the experiment is repeated but with the surface of the bulb blackened, the temperature rises at $0.0186\ \mathrm{K\,s^{-1}}$.
a) How do you explain the difference of heating rates?
b) If glass and water are opaque in the infra-red, estimate the proportion of the power delivered to the bulb which is converted into light.

6.5 Explain what is meant by *spectral emissive power*. Why is it necessary to define such a quantity?

In the visible part of the spectrum, the spectral emissive power of the tungsten filament of an electric lamp varies approximately as $\exp-(hc/\lambda kT)$ where h is the Planck constant, c the speed of light, λ the wavelength of the light, k the Boltzmann constant and T the temperature. Estimate the ratio of the intensities in wavelength bands 10 nm wide centred (a) in the red ($\lambda \approx 670$ nm) and (b) in the blue ($\lambda \approx 450$ nm) when $T = 3000$ K.
[$h = 6.63 \times 10^{-34}\ \mathrm{J\,s}$; $k = 1.38 \times 10^{-23}\ \mathrm{J\,K^{-1}}$.]

Prévost's theory of exchanges. Equal temperature enclosures; equilibrium radiation; black bodies

6.6 What is *equilibrium radiation*? Why is it also called *black body radiation*?

Explain why, when looking into the depths of a fire, it is often difficult to distinguish the shapes of objects in it.

Influence of surfaces; spectral absorptivity and emissivity; Kirchhoff's law

6.7 If clean quartz is heated in a hot flame it is possible to 'work' it like glass; but if the quartz has been contaminated, by grease from the fingers, for example, it glows brightly when held in the flame but does not become soft enough to work. Explain these observations.

The Stefan–Boltzmann law; energy exchange by radiation; Newton's law for small temperature differences. Energy density of radiation.

6.8 The filament of a 100 W lamp has a radius of 12 μm and is 0.3 m long. Estimate the temperature at which it operates.
What assumptions are you making?

6.9 The road surface of The Mound in Edinburgh is electrically heated to prevent icing in winter. If there were negligible warming from the air on a very still clear night, roughly how much power would have to be supplied to each square metre to prevent the surface temperature falling below freezing point? (Assume the average emissivity of the surface is 0.7.)

How much power would be needed to melt snow falling at the rate of 2 mm h^{-1}?
(Precipitation rates are measured in mm of equivalent water per hour.)
[Latent heat of melting of ice = 333 J g^{-1}. Radiation temperature of interstellar space ≈ 4 K.]

6.10 A long rod of niobium, 4 mm in diameter, is being annealed in a vacuum furnace at 2100 K.
a) If the surroundings are at room temperature, what heating rate is required?

The rod is now surrounded by a radiation shield in the form of a thin-walled cylinder of radius 10 mm. If T is the temperature of the cylinder, and using symbols for the various quantities concerned, write expressions for
b) the rate at which the cylinder gains heat from the rod,
c) the rate at which the cylinder loses heat to the surroundings. (Beware the culpable mistake; see section 6.7.1.)
In the steady state, the cylinder must have no net energy exchange. Hence
d) derive an expression for the temperature of the cylinder,
e) substitute values to find the temperature in kelvins,
f) find the heating rate which is now required to keep the niobium rod at 2100 K.
[Assume all surfaces are black.]

6.11 The inflammation resulting from arthritis in a knee produces a skin temperature rise of 2 °C.
a) How much extra energy is radiated per second by each square centimetre of affected area? (Assume that skin is black in the far infra-red.)
b) If this extra energy is radiated approximately equally in all directions, how much extra energy passes through each square centimetre at a distance of 2 m?
c) How much extra energy from each square centimetre of affected area enters a thermographic camera 2 m away with an entrance aperture 6 cm in diameter?

6.12 In experiments at very low temperatures it is necessary to shield the cold parts of the apparatus from radiation from objects at room temperature, which would cause significant heating.

In a low temperature apparatus, a copper plate of area 100 mm^2 is exposed to room temperature radiation from a surface of area 100 mm^2 0.5 m away. The copper plate is supported in a vacuum on a german silver wire, the far end of which makes contact with a helium bath at 1.0 K. If all surfaces are blackened, and the thermal resistance of the german silver wire is 0.2 K μW^{-1}, estimate the equilibrium temperature of the copper plate.

6.13 The radius of the sun is 6.96×10^8 m and its surface temperature is 6000 K.
a) Calculate the *solar constant*: the power passing through each square metre at a distance equal to the radius of the earth's orbit.
b) Calculate the energy intercepted by the earth assuming all incident energy is absorbed.
c) Calculate the equilibrium temperature of the earth assuming it behaves as a black body and reaches a uniform temperature throughout.
d) How would your answer to (c) be affected if you took into account the fact that the earth is not black but has a mean emissivity of 0.64?
[Radius of earth's orbit = 1.50×10^{11} m.
Radius of earth = 6370 km.]

6.14 Show that the equilibrium temperature T of a comet at a distance R from the sun is given approximately by

$$T = T_s\left(\frac{r_s}{2R}\right)^{\frac{1}{2}}$$

where r_s and T_s are the radius and surface temperature of the sun.

A comet approaching the sun from a great distance consists of solid hydrogen. When it warms to about 14 K the vapour pressure becomes sufficiently large for the hydrogen to sublime at an appreciable rate. The gas released forms the comet's tail. Roughly how far is the comet from the sun when it develops a significant tail? Take the surface temperature of the sun as 6000 K; give your answer in sun's radii, and make clear any assumptions you make.

6.15 Design a 60 W 12 V car headlamp bulb:
a) Write an expression for the power radiated in terms of the Stefan–Boltzmann constant and the radius, length and temperature of the filament.
b) Write an expression for the electrical power drawn from the battery in terms of the electrical resistivity, the dimensions of the filament, and the voltage of the supply.

The filament is to operate at 3000 K at which temperature the resistivity of tungsten is $7 \times 10^{-7}\ \Omega$ m.
c) What length and radius must the filament have?
d) What assumptions have you made in your calculations?

6.16 A solar furnace consists of a parabolic mirror which focuses the sun's rays onto the object to be heated. Discuss the factors which determine how hot the object becomes. What is the highest temperature which could be attained?

6.17 A solid copper sphere of radius 20 mm is suspended in a vacuum within a container whose walls are at 20 °C. When the sphere is at 120 °C its cooling rate is 0.11 K min^{-1}. What will be the cooling rate of a 30 mm radius solid copper sphere when its temperature is 200 °C?

6.18 A solid copper cylinder, 10 mm in radius and 40 mm long is supported in a vacuum. The walls of the container are cooled by liquid nitrogen to 77 K. What is the initial cooling rate of the copper if its initial temperature is 20 °C? (Take all surfaces as black.)

What would be the rate for a cylinder twice as long? [For copper, specific heat capacity at 20 °C = 385 $\text{J K}^{-1}\ \text{kg}^{-1}$ and density = $8.93 \times 10^3\ \text{kg m}^{-3}$.]

6.19 In a low temperature calorimetric experiment a specimen of surface area 1000 mm^2 is suspended in a container in a vacuum. To reduce exchange of heat the temperature of the walls of the container is kept close to that of the specimen. If the maximum temperature difference which can develop is 0.5 K and the temperature of the experiment is 77 K, what is the maximum power exchanged between specimen and surroundings, assuming all surfaces to be black?

6.20 Estimate the total radiant energy present in a room 4 m × 8 m × 3 m.

Wien's displacement law; colour of hot glowing bodies

6.21 What are the wavelengths of maximum spectral emissive power of black surfaces at

a) 4 K (normal boiling point of helium, temperature of interstellar space),
b) 290 K (room temperature),
c) 1000 K (red hot poker),
d) 3000 K (lamp filament),
e) 10^8 K (centre of the sun or core of a hydrogen bomb)?

6.22 A 1 kW electric heater element is 300 mm long and has a radius of 8 mm. Use your knowledge of the colour with which such an element glows to obtain an estimate of the Stefan–Boltzmann constant. (Use table 6.1 on page 100.)

What would you think is the greatest source of error in trying to estimate Stefan's constant in this way?

6.23 A person sitting 2 m from a domestic coal fire sees about 100 cm^2 of coal at yellow heat. Estimate the power incident on his face.

Planck radiation law and birth of quantum theory

6.24 What is the quantum of oscillation of
a) a pendulum of period 1 s,
b) the tuned circuit of a television receiver (600 MHz),
c) the vibration of an hydrogen chloride molecule corresponding to the spectral line at 3.5 μm,
d) the 154 pm copper K-α X-ray line?
Give your answers in joules, electronvolts and (equivalent) kelvins.

6.25 γ for hydrogen has the value 5/3 at 50 K and the value 7/5 at 300 K.
a) What can you deduce about the number of effective degrees of freedom at these temperatures?
b) Why does the number change?
c) Estimate the order of magnitude of the quantum of energy of the motions which are 'frozen out' as the temperature is lowered from 300 to 50 K.

7 Heat into Work

7.1 THINGS THAT DO NOT HAPPEN

The consequences of the first law of thermodynamics (page 18) may be summarized by the statements that (a) heat is a form of energy, and (b) energy is conserved in thermal processes. In chapter 2, when we were discussing heat as a form of energy, we considered processes in which some form of work was entirely converted into heat. For example, we heated a gas by doing electrical work on a resistor. The whole of the electrical work was dissipated as heat; there was no stored electrical energy which could be extracted afterwards. Another example is work against friction. Such processes result in the *complete* conversion of work into heat. But what about the reverse, the conversion of heat into work? A steam engine uses a source of heat to produce work. Is it possible to bring about the complete conversion of heat into work? Common experience tells us that it is not. If it were possible, we could power a boat to cross the Atlantic by simply absorbing heat from the sea, and using this energy to drive the propellers. We would need no fuel and travel would be cheap! Similarly, we cannot reverse our electrical heating: we cannot make the resistor reabsorb the heat from the gas and produce electrical energy at its terminals. Such processes simply do not occur. One of the well known statements of the *second law of thermodynamics*. *The Kelvin statement*:

No process is possible whose sole result is the complete conversion of heat into work.

The first law forbids perpetual motion of the first kind: a machine that creates its own energy. The second law forbids *perpetual motion of the second kind*: a machine which just absorbs heat and produces work (figure 7.1).

What about the isothermal expansion of an ideal gas? you may ask. Does that violate the Kelvin statement of the second law? The internal energy of an ideal gas only depends on temperature (Joule's law). When the gas expands, it does mechanical work on the piston it is pushing back. At the same time, if the expansion is isothermal, it must absorb heat so that the internal energy may stay constant, which it must if the temperature does not change (figure 7.2):

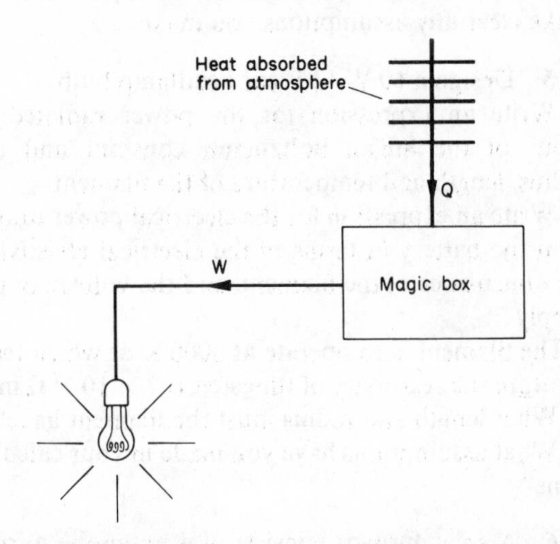

Figure 7.1 Perpetual motion of the second kind. A machine which produces complete conversion of heat into work violates the Kelvin statement of the second law of thermodynamics.

$$\Delta U = Q + W = 0$$

so

$$W = -Q$$

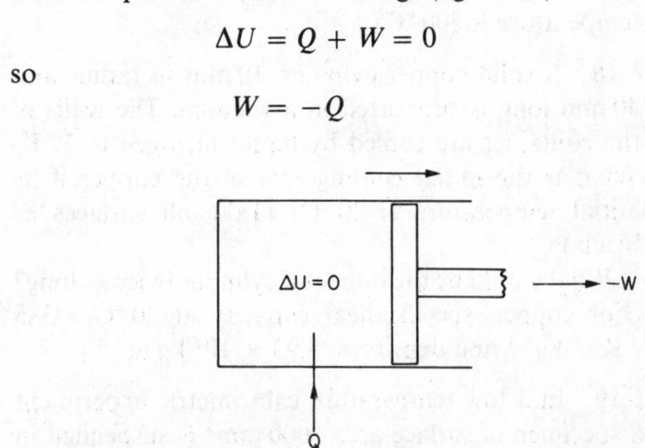

Figure 7.2 Isothermal expansion of an ideal gas. At constant temperature, U is constant, so that when the gas expands, the work done is equal to the heat absorbed.

where W is the work done *on* the gas (using our usual sign convention), and Q is the heat absorbed. Therefore the work done *by* the gas is just Q and we have complete conversion of heat into work! This does not violate the Kelvin statement, however, because the state of the gas is different at the end from what it was at the beginning. The conversion of heat to work is not the *sole* effect. We could not use the gas to make a machine which would convert heat into work continuously because, at some stage, we would have to recompress the gas to start again from the original state, and there is no way we could do this without generating at least as much heat as we converted in the first place.

There are other processes which conserve energy but simply do not happen: heat does not spontaneously (of its own accord) flow from colder to hotter. If we put two bodies with heat capacities C_1 and C_2 at temperatures T_1 and T_2 into thermal contact, we know that heat flows until they reach a common temperature T_f. If the heat capacities are constant over the relevant temperature range, T_f is given by

$$T_f(C_1 + C_2) = T_1C_1 + T_2C_2$$

(We write this equation by conserving total energy:

$$\Delta U_1 + \Delta U_2 = 0$$

with

$$\Delta U_1 = Q_1 = C_1(T_f - T_1),$$

$$\Delta U_2 = Q_2 = C_2(T_f - T_2).)$$

But we could also conserve energy if the temperatures *diverged*: if the hot body got hotter and the cold colder. This would involve heat flowing (spontaneously) from cold to hot. Such things do not happen. There is a one-way aspect to heat flow which has nothing to do with conservation of energy. A second well known statement of the second law is based on this fact.

The Clausius statement:*

No process is possible whose sole result is the transfer of heat from a colder to a hotter body.

This statement is not violated by a refrigerator which has to be supplied with energy to make it operate.

The two statements we have given of the second law look quite different, but it can be shown that each implies the other so that their consequences are identical. These very simple statements contain all the information which is needed to deal quantitatively with conversion of heat into work and with the one-way aspect of natural heat flow. We will not follow through the arguments here (which are the core of thermodynamics), but we will look at some of the ideas and results which come out of them.

* Rudolf Clausius, 1822–1888

7.2 HEAT ENGINES

According to the second law, heat wants to flow from hot to cold. We cannot achieve complete conversion of heat into work; but we might expect to be able to *intercept* heat as it flows from hot to cold with some machine which will convert some of the heat into work. It is rather like water flowing downhill: if it just tumbles down the hillside, the potential energy is wasted in turbulence, viscous motion and noise; but if we intercept the flow and divert it through a turbine, we can convert some of the potential energy to useful work. A machine which intercepts the flow of heat and converts some of the heat into work is called a *heat engine* (figure 7.3). To be a useful device, the heat engine must be able to operate continuously, absorbing heat at a high temperature, rejecting heat at a colder, and producing work. It follows that the processes which take place inside the engine must not result in permanent change. This means that the engine must work in a *cycle*: a series of processes at the end of which it arrives back in its initial state. The bodies with which it exchanges heat during the cycle are

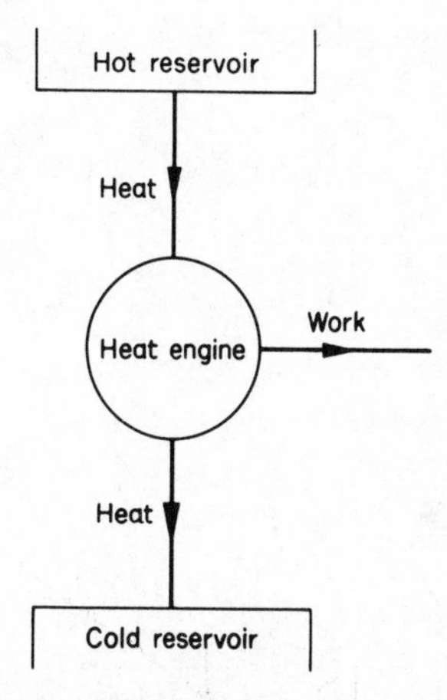

Figure 7.3 A heat engine. The engine intercepts the flow of heat and converts some of the heat into work.

called *reservoirs*; and the material the engine operates with is called the *working substance*. For example, the operation of a steam engine may be represented in simplified form by a cycle as follows (figure 7.4).

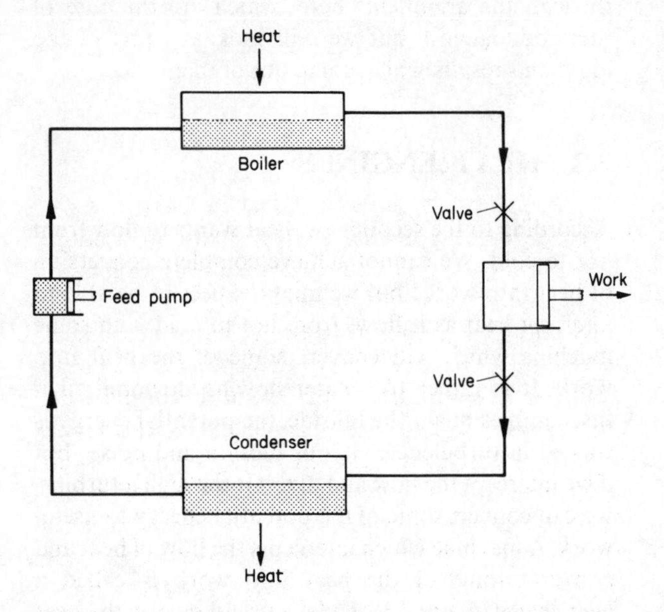

Figure 7.4 A steam engine

The working substance is water which is vaporized in the boiler to produce steam at high pressure. The steam forces back a piston, doing external work, and is then recondensed in a condenser and returned to the boiler. During the condensation the pressure falls and the piston is returned to its original position. The boiler is the hot reservoir, the condenser the cold reservoir. At the end of the cycle, the engine is in its original state, but heat has been absorbed from the boiler, rejected to the condenser, and net work has been done. Figure 7.5 shows a water analogue of a heat engine.

We define the *efficiency* η of a heat engine as the proportion of the heat absorbed which is turned into work. If, during one cycle, the engine absorbs heat Q_1 from the hot reservoir, rejects Q_2 at the cold reservoir and does work W (figure 7.6), the efficiency is

$$\eta = W/Q_1 \tag{7.1}$$

Since the cycle returns the engine to its original state, its energy at the end must be the same as it was at the beginning; that is, $\Delta U = 0$. But during the cycle, work W has been done by the engine and net heat $Q_1 - Q_2$ absorbed. Hence,

$$\Delta U = Q_1 - Q_2 - W = 0$$

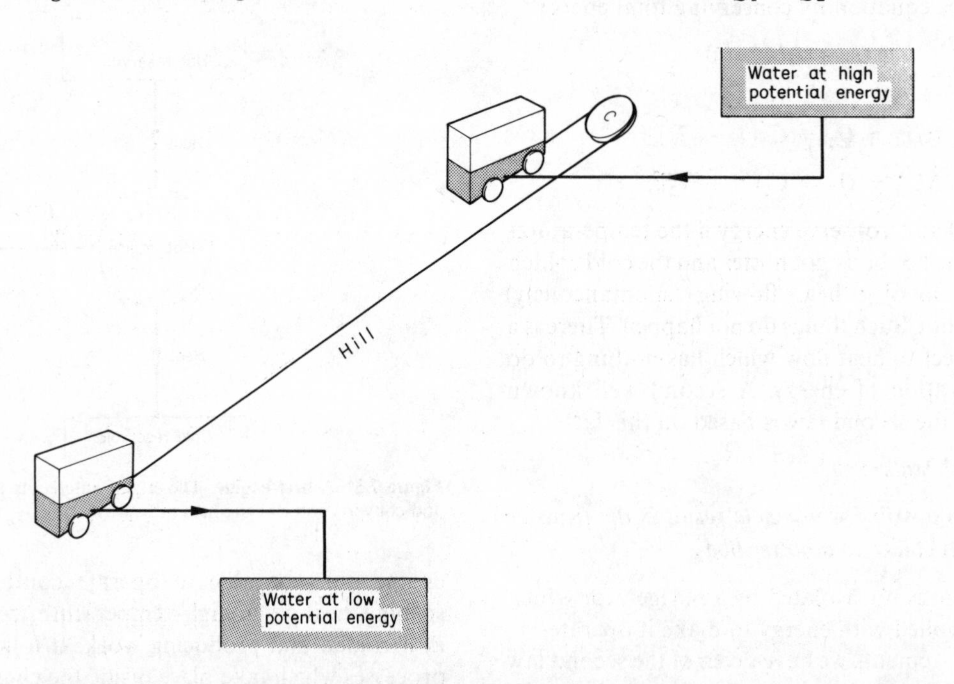

Figure 7.5 The Lynton and Lynmouth Cliff Railway in Devon operates as a water analogue of a heat engine. In the cycle, first one carriage descends and then the other, returning the first to the top. The railway is driven by filling a tank in the upper carriage with water (at high potential energy, equivalent to heat at a high temperature) until it becomes heavier than the lower carriage and begins to descend. At the bottom, the water is discharged (now at low potential energy, equivalent to rejection of heat at a low temperature).

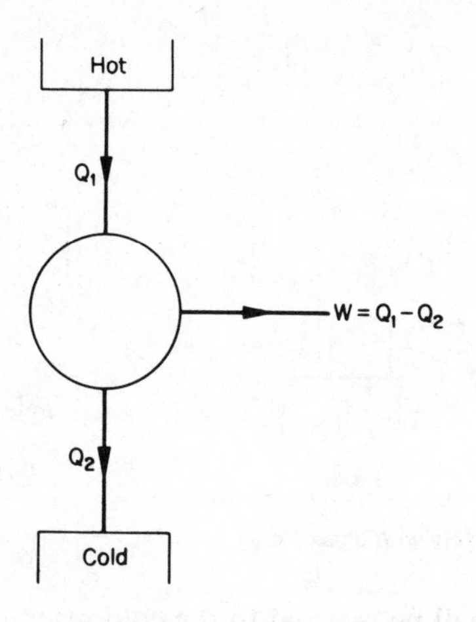

Figure 7.6 The energy changes in one cycle of a heat engine

so

$$W = Q_1 - Q_2 \qquad (7.2)$$

Substituting in *7.1*, we get an alternative formula for η,

$$\eta = 1 - Q_2/Q_1 \qquad (7.3)$$

7.3 MAXIMIZING THE EFFICIENCY

Now the second law tells us that $\eta = 1$ is impossible (that would violate the Kelvin statement), but it is of enormous practical importance to discover just how high the efficiency can be made. There are two questions we need to answer:

i) What general conditions must the engine satisfy if the efficiency is to be as high as possible?
ii) Are particular engines and working substances better than others? Would it be better to operate a steam engine on benzene instead of water, for example?

It was the work of Sadi Carnot (1796–1832) in seeking answers to these questions which laid the foundations of thermodynamics.

We can guess at the answers to the first question: what general conditions must the engine satisfy? Firstly, the engine must be *mechanically reversible*: there must be no friction or hysteresis (as in the magnetization of iron) or turbulence to waste energy. Provided that these are absent, then a process may be exactly reversed by making an *infinitesimally small* adjustment to the conditions. For example, if we are pulling a frictionless trolley steadily up an incline, our pull just balances gravity. As there is no friction, the tension would be unchanged if the trolley were moving steadily *down* the incline. Therefore, *all* the work which is stored as potential energy when the trolley is raised, is extracted again as it is lowered. Hence the term *reversible*. If, on the other hand, friction were present, then, for the trolley to move downhill, the tension in the string would have to be less than while it was going up (by twice the frictional force). This is a finite change in conditions. As the trolley ascends, some work is lost in friction. As it descends, some potential energy is lost in friction. Thus, less energy is extracted than was originally put in: the process is irreversible.

Secondly, there is a thermal equivalent of mechanical reversibility. We get work by intercepting flow of heat from hot to cold. To obtain the most work, we do not want the heat to flow from a higher to a lower temperature without being intercepted. Think of the water analogue: if the water runs half way down the hill before we collect it to make it drive the turbine, we shall have wasted half its potential energy and will get less work out of it. In the thermal case, we must never have heat flowing down a temperature gradient: the temperature differences through which heat flows must be negligible. In this case, the thermal processes are reversible in the same sense as the mechanical ones were: an infinitesimally small adjustment (of temperature) will reverse a heat flow. The system is then said to be *thermally reversible*. If a process is both mechanically and thermally reversible, it is simply said to be (thermodynamically) reversible. We may therefore sum up these ideas on efficiency by the statement: *for maximum efficiency, a heat engine must be reversible.*

7.4 CARNOT ENGINES

We now turn our attention to the second question: are particular cycles or working substances better than others?

The simplest possible reversible cycle is the *Carnot cycle* in which heat is exchanged with two reservoirs only. Any system can be made to perform a Carnot cycle, but we will illustrate it here with a gas.

Suppose the gas is in a cylinder with a frictionless piston. Let the temperatures of the hot and cold reservoirs be T_1 and T_2 respectively. The cycle is illustrated in figures 7.7 and 7.8. We start with the gas

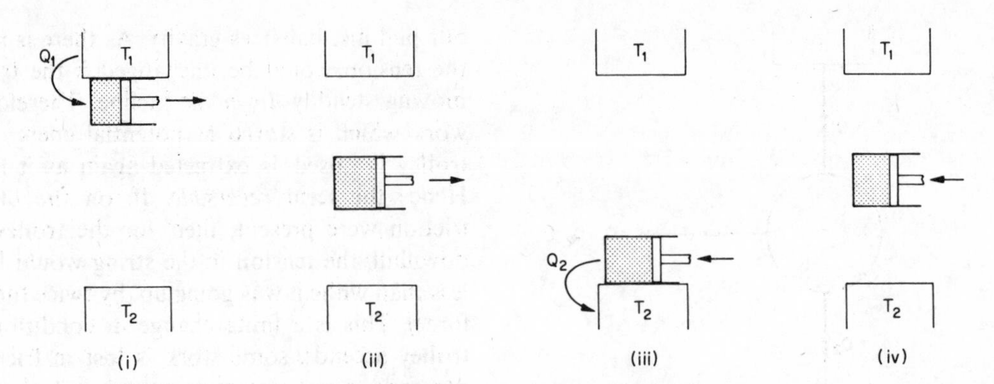

Figure 7.7 A Carnot cycle with a gas

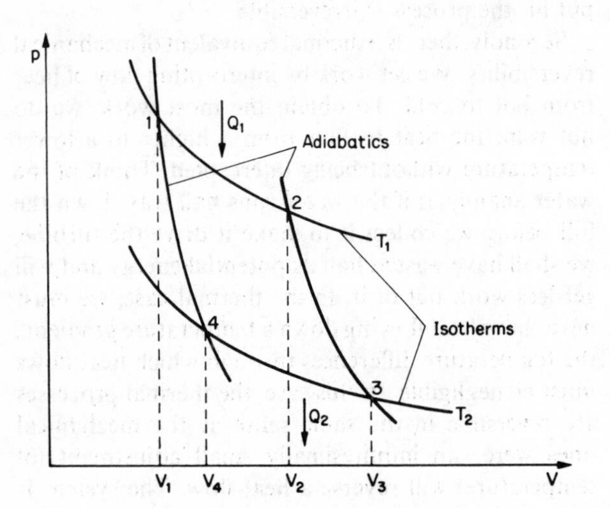

Figure 7.8 A gas Carnot cycle on a *p-V* plot

in thermal contact with the reservoir at T_1 and with volume V_1. The cycle has four parts, all of which we assume to be carried out reversibly.

i) Keeping thermal contact with the hot reservoir, let the gas expand to V_2. In this process, work will be done *by* the gas on the surroundings (via the piston) and, as the change is isothermal, heat Q_1 will be absorbed (so as to prevent the temperature from falling).
ii) Thermally isolate the gas and expand further. Again, work will be done *by* the gas, but this time the temperature will fall because the change is adiabatic. When the temperature has fallen to T_2, stop expanding and call the volume reached V_3.
iii) Place the gas in thermal contact with the cold reservoir at T_2 and compress from V_3 to V_4. During this process, work will be done *on* the gas and heat Q_2 will be rejected to the cold reservoir (so as to prevent the temperature from rising).
iv) Thermally isolate the gas and compress further until the temperature rises to T_1. Again, work will be done *on* the gas. By choosing V_4 correctly, we can arrange that the volume has reduced to V_1 by the time the temperature reaches T_1. Thus we have completed the cycle, which, on the *p-V* diagram, consists of the intersection of two isotherms ($1 \rightarrow 2$ and $3 \rightarrow 4$) and two adiabatics ($2 \rightarrow 3$ and $4 \rightarrow 1$). All processes are reversible: there is no mechanical friction and at no time does heat flow across a temperature difference. (The system is at T_2 before we make thermal contact with the cold reservoir, etc.)

The total work done *by* the gas in the cycle is

$$W = \oint p \, \mathrm{d}V$$

where the symbol $\oint$ means that we integrate all the way around the cycle back to the starting point. This integral consists of four parts:

$$\oint p \, \mathrm{d}V = \int_1^2 p \, \mathrm{d}V + \int_2^3 p \, \mathrm{d}V + \int_3^4 p \, \mathrm{d}V + \int_4^1 p \, \mathrm{d}V$$

The first contribution is the area below the part of the cycle from $1 \rightarrow 2$. The second is the area below $2 \rightarrow 3$. But the third is *minus* the area under $3 \rightarrow 4$ because we integrate from the larger to the smaller volume. (We are doing work *on* the gas as we compress it, so the work done *by* the gas is negative.) Similarly, the last contribution is minus the area under $4 \rightarrow 1$. All that is left is the area *inside* the cycle $1 \rightarrow 2 \rightarrow 3 \rightarrow 4 \rightarrow 1$. In one cycle, therefore, *the net work done by the engine is numerically equal to the area enclosed by the cycle in the p-V diagram.* (Remember that $p \times V$ has the dimensions of energy.)

The theory of thermodynamics uses the second law to show that

i) all reversible engines (i.e. Carnot engines) operating between the same reservoirs are equally efficient; and
ii) no engine can be more efficient than a reversible engine operating between the same reservoirs.

The Carnot engine is therefore the ideal engine. It represents the best that could possibly be achieved. Remember that, although we have illustrated Carnot cycles by referring to a gas, *any* system can be made to follow a Carnot cycle and the results are quite general.

Now the fact that all reversible engines operating between the same reservoirs are equally efficient answers the second of the questions we asked on page 107. *The efficiency of a reversible engine does not depend on the working substance or on the nature of the engine, but only on the reservoirs.* From *7.3*, the formula for efficiency, we see that this means that Q_1/Q_2, the ratio of the heats exchanged at the reservoirs, also can only depend on the reservoirs. But the only property of the reservoirs which comes in, is their temperature. *The amount of heat exchanged therefore gives a measure of temperature which is independent of the properties of any particular substance.* This fact is used to define thermodynamic temperature.

The ratio of the thermodynamic temperatures of two reservoirs is defined to be equal to the ratio of the amounts of heat exchanged at the reservoirs by a reversible engine operating between them.

$$T_1/T_2 = Q_1/Q_2 \qquad (7.4)$$

Since the definition is in terms of a ratio, it does not fix the size of the unit of thermodynamic temperature which is left as a matter for choice. The size was originally chosen to give 100 units between ice and steam points so as to coincide with the perfect gas centigrade scale (apart from the shift of zero); but the modern definition sets the size of the unit by fixing the value of the thermodynamic temperature of the triple point of water as 273.16 K (see section 1.5).

In terms of thermodynamic temperature, the efficiency of an ideal heat engine operating between two reservoirs becomes

$$\eta = 1 - T_2/T_1 \qquad (7.5)$$

7.5 REAL HEAT ENGINES

Real heat engines—steam engines, internal combustion engines, turbines, and so on—are far from being thermodynamically reversible. They are mechanically irreversible because there is friction, and, often, parts of the cycle are violent (like the explosion in a petrol engine) and so involve turbulence and other processes which dissipate useful energy. They are also thermally irreversible because the cycles involve flow of heat across temperature differences so, again, some of the potential for doing useful work is lost. How can we use the results we have given for ideal heat engines to help us estimate the efficiences of real engines?

To make a very rough estimate of the maximum possible efficiency of a heat engine is simple. For a reversible engine operating between two reservoirs, we know that the efficiency increases as the ratio of the temperatures of the reservoirs increases (equation *7.5*). We also know that no real engine can be more efficient than a reversible engine operating between the same temperatures. Therefore, *the efficiency of a real heat engine must certainly be less than that of a reversible engine operating between two reservoirs whose temperatures equal the highest and lowest temperatures involved in the cycle of the real engine.* We choose the highest and lowest temperatures because this gives a maximum value for the efficiency of the ideal cycle.

Making this sort of extreme comparison, we can see immediately why the efficiency of simple steam engines is low. (The equivalent cycle was described on page 106.) The highest temperature involved is that of the boiler where the steam is produced at high pressure. If the engine uses steam at 10 atm pressure, the boiling point will be about 180 °C ≈ 453 K. The condensation might be at 60 °C ≈ 333 K corresponding to a vapour pressure of 0.2 atm. Then a reversible engine operating between reservoirs at these temperatures would have an efficiency

$$\eta = 1 - 333/453 = 26\%.$$

The efficiency of the real engine, because of its irreversibilities, would be much less than this—a few per cent only. Modern steam engines are better, partly because of the use of higher pressures (which increases T_1) and partly because technical improvements have eliminated some of the more wasteful aspects of the early engines. The most efficient use of steam power is in the modern steam turbine where high pressure water vapour is heated to over 600 °C before entering the turbine. Efficiencies greater than 40% are achieved.

Comparison of real heat engines with a Carnot engine gives only a very rough indication of maximum efficiency. A better estimate can be made by inventing a reversible cyclic process which is similar to the cycle of the real engine, and analyzing that in detail. The cycle used for the petrol engine is shown in figure 7.9. It is known as the Otto cycle. The working substance is taken to be air. The cycle has six parts.

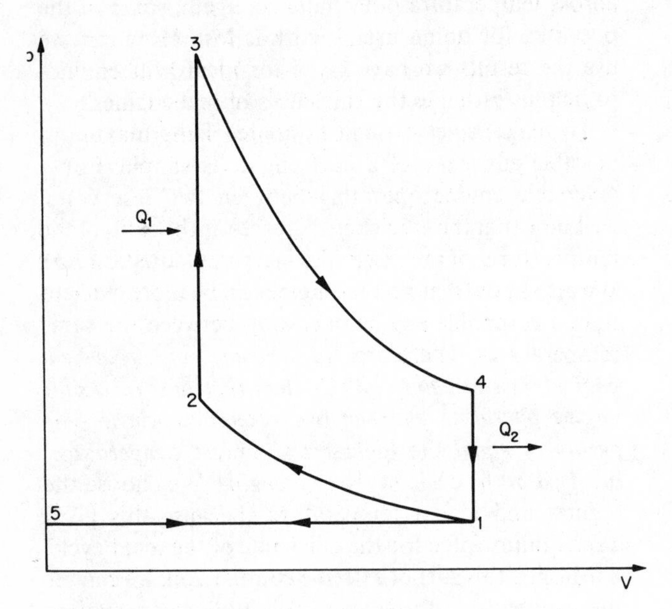

Figure 7.9 The Otto cycle

1) *Intake stroke*, 5 → 1. Air is drawn into the cylinder at constant pressure.
2) *Compression stroke*, 1 → 2. The air is compressed adiabatically.
3) *Explosion*, 2 → 3. When the air is fully compressed, heat is added and the temperature and pressure rise. The explosion is rapid, so there is no change in volume while the heat is being added; but, for the idealized cycle, we suppose that the addition of heat is reversible.
4) *Power stroke*, 3 → 4. The hot air expands adiabatically, doing work.
5) *Valve exhaust*, 4 → 1. In the real engine, when the exhaust valve opens, the gases which are still at high pressure rush out and the pressure drops to atmospheric. In the idealized cycle, this is represented by the air losing heat reversibly until its temperature and pressure drop to the same values they had when the air was drawn in.
6) *Exhaust stroke*, 1 → 5. The air is forced out of the cylinder as the piston moves up in preparation for drawing in the air for the next cycle.

The analysis of this cycle gives for the efficiency

$$\eta = 1 - 1/r^{\gamma-1} \qquad (7.6)$$

where γ is the ratio of the principal heat capacities of air, and r is the *compression ratio*, the ratio of the maximum to minimum volumes. The efficiency increases as the compression ratio is increased, but it cannot be made larger indefinitely because, eventually, the rise in temperature during compression becomes large enough to ignite the petrol–air mixture before compression is complete. Typically, r might be 10 and, taking $\gamma = 1.4$, this gives an efficiency of 60%. In the real engine, the efficiency probably only reaches about half this value.

7.6 VAPOUR PRESSURE

So far, all this discussion of the second law must seem very engineering oriented, very concerned with the efficiency of engines. Historically, this is where the early development of thermodynamics occurred. It was only later that the importance of its laws to other branches of science was realized. As an example of a quite different application of thermodynamics, we will derive a very important result about change of phase. The basic result applies to any simple phase change like solid to liquid (melting), liquid to vapour (evaporation and boiling), solid to vapour (sublimation). To make the argument easy to follow, we will explain it for the second case, evaporation.

We have already discussed what happens when a vapour is compressed at constant temperature (page 65): (i) the pressure rises as the volume is reduced until liquid starts to condense out; (ii) further compression then causes more liquid to condense, the pressure remaining constant at the saturated vapour pressure, until (iii) when all the vapour is condensed to liquid, further compression causes the pressure to rise rapidly because liquids are not very compressible. Typical curves are sketched in figure 7.10 for two temperatures which are taken to be close together, T and $T + \mathrm{d}T$. Now we can carry out a Carnot cycle using these two neighbouring isotherms. Starting at 1, where the substance is just all liquid and the temperature is $T + \mathrm{d}T$, the cycle is as follows:

1 → 2 Isothermal expansion during which the liquid

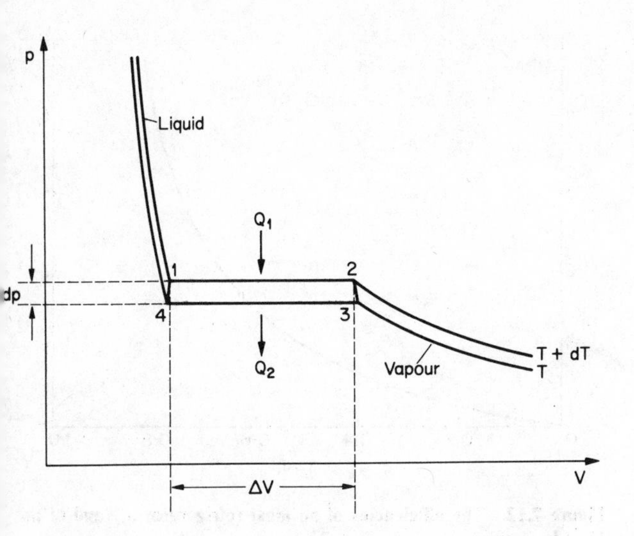

Figure 7.10 Derivation of the Clausius–Clapeyron equation

evaporates. The evaporation requires latent heat, so heat Q_1 is absorbed.
$2 \rightarrow 3$ A small adiabatic expansion during which the temperature falls by the small amount $\mathrm{d}T$.
$3 \rightarrow 4$ An isothermal compression to the point where all the vapour is condensed. Latent heat Q_2 is rejected.
$4 \rightarrow 1$ A small adiabatic compression during which the temperature rises by $\mathrm{d}T$ back to $T + \mathrm{d}T$.

Now the pressure at the horizontal parts of the curves is just the vapour pressure: liquid and vapour are both present. Suppose the increase in vapour pressure in going from T to $T + \mathrm{d}T$ is $\mathrm{d}p$ (figure 7.10). Then the work done in the cycle is the area enclosed:

$$W = \mathrm{d}p\,\Delta V$$

where ΔV is the mean difference between the volumes of the vapour and the liquid. Equations *7.1* and *7.5* for the efficiency of the heat engine using this cycle give

$$\eta = W/Q_1 = 1 - T_2/T_1$$

Substituting for W, and putting $Q_1 = L$ (the latent heat) and $T_1 - T_2 = \mathrm{d}T$, we obtain

$$\mathrm{d}p\,\Delta V/L = \mathrm{d}T/T$$

Rearranging, we have an expression for the rate of change of vapour pressure with temperature:

$$\frac{\mathrm{d}p}{\mathrm{d}T} = \frac{L}{T\Delta V} \qquad (7.7)$$

The same argument can be applied to any similar change of phase and would lead to the same result with L as the appropriate latent heat and ΔV the appropriate change in volume in passing from one phase to the other. Equation *7.7* is a very important result known as the *Clausius–Clapeyron equation.*

We will now use the Clausius–Clapeyron equation to derive an approximate formula for how vapour pressure depends on temperature. The Clausius–Clapeyron equation is exact, but it is a *differential equation*: it tells us how rapidly pressure changes with temperature. To obtain a formula which connects vapour pressure and temperature explicitly, the Clausius–Clapeyron equation must be integrated. We can only do this by making approximations.

i) Usually the volume of a vapour will be much greater than the volume of the same amount of substance as liquid. Therefore we may put

$$\Delta V = V_{\text{vap}} - V_{\text{liq}} \approx V_{\text{vap}}$$

ii) We suppose that the vapour behaves like an ideal gas. Then, if we apply the Clausius–Clapeyron equation to one mole,

$$\Delta V = V_m = RT/p$$

iii) We assume that the latent heat L is independent of temperature.

Then the Clausius–Clapeyron equation, applied to one mole, becomes

$$\frac{\mathrm{d}p}{p} = \frac{L_m}{R}\frac{\mathrm{d}T}{T^2}$$

Integrating this,

$$\ln p = -\frac{L_m}{RT} + a$$

where a is the constant of integration. Then,

$$\mathrm{e}^{\ln p} = p = \mathrm{e}^{-L_m/RT + a}$$

which may be written

$$p = p_0\,\mathrm{e}^{-L_m/RT} \qquad (7.8)$$

where we have put $\mathrm{e}^a = p_0$.
This result for vapour pressure in terms of temperature and latent heat is similar to equation *5.12* of chapter 5 (page 81); but here we have derived it without making any microscopic assumptions about the energetics of evaporation, nor have we had to use the Boltzmann factor. This is a good illustration of the power of thermodynamics.

7.7 REFRIGERATORS AND HEAT PUMPS

The Carnot cycle is reversible, so it may be run backwards by supplying mechanical work. The result will be to transfer heat in the unnatural direction, from cold to hot (figure 7.11). Such an arrangement is

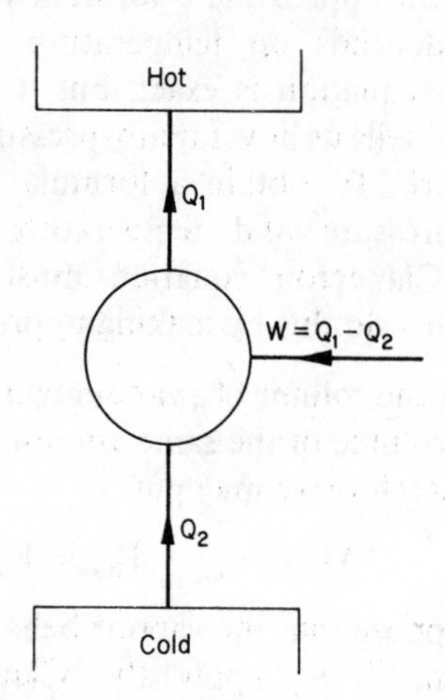

Figure 7.11 A heat engine driven backwards transfers heat in the unnatural direction, from cold to hot.

called a refrigerator or a heat pump, depending on whether one is interested in the extraction of heat from the colder temperature or the delivery of heat to the hotter.

With refrigerators, we are interested in the extraction of heat from the low temperature. We may define an efficiency or *figure of merit* for a refrigerator

$$\eta_r = \frac{\text{heat absorbed at low temperature}}{\text{work required}}$$

$$= \frac{Q_2}{W} = \frac{Q_2}{Q_1 - Q_2}$$

Just as it makes the best possible heat engine, in the same way a Carnot engine, because it has no irreversibilities, would make the best possible refrigerator. Then substituting for the Q's in terms of temperature, we get

$$\eta_{rC} = \frac{T_2}{T_1 - T_2} \qquad (7.9)$$

η_{rC} is shown in figure 7.12. As long as T_2 is at least $T_1/2$, more heat is extracted than work required; but as T_2 becomes much smaller than T_1, the work required to extract a given amount of heat rises rapidly.

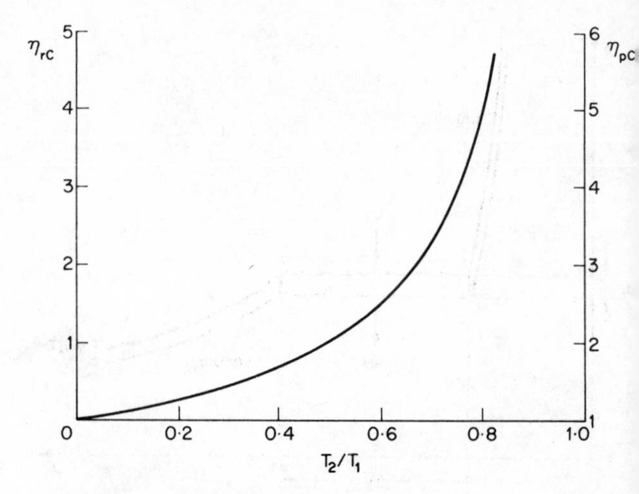

Figure 7.12 The efficiencies of an ideal refrigerator η_{rC} and of an ideal heat pump η_{pC}

For a domestic refrigerator, the cold box might be at about $T_2 = 260$ K (≈ -13 °C), and the temperature of the cooling fins through which the heat is rejected to the surroundings would be somewhat above room temperature, say at $T_1 = 300$ K (≈ 27 °C). Then an ideal domestic refrigerator would have $\eta_r = 6.5$, so that 1 W of electrical power would extract heat at the rate of 6.5 W. Of course, the efficiency of a real domestic refrigerator is much lower.

The figures work out very differently when we examine what happens when a physicist wants to do experiments at very low temperatures. An ideal refrigerator extracting heat at the rate of 5 W at 1 K and rejecting heat at room temperature would consume nearly 1.5 kW! and, again, the performance of any real refrigerator is many times worse. This shows why it becomes increasingly difficult to reach very low temperatures.

With heat pumps, we are interested in the heat delivered at the higher temperature, so the figure of merit becomes

$$\eta_p = \frac{\text{heat delivered at high temperature}}{\text{work required}}$$

$$= \frac{Q_1}{W} = \frac{Q_1}{Q_1 - Q_2}$$

Using a Carnot engine as an ideal heat pump,

$$\eta_{pC} = \frac{T_1}{T_1 - T_2} \qquad (7.10)$$

η_{pC} is also shown in figure 7.12.

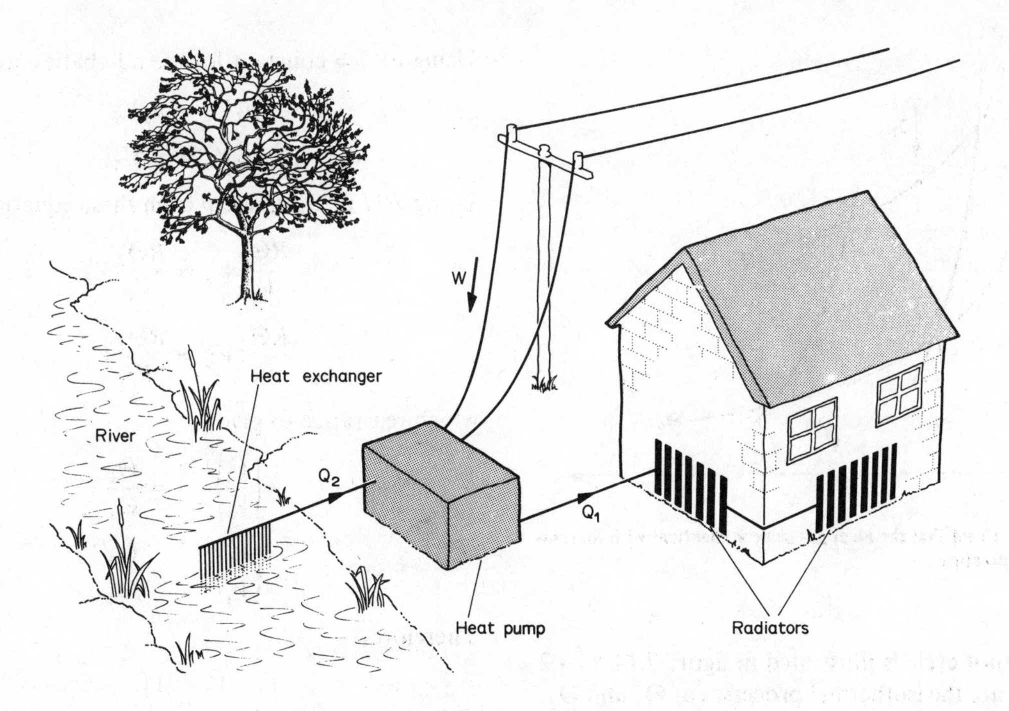

Figure 7.13 Using a heat pump to heat a building

Now the interesting thing about heat pumps is that you always get more heat out than you put energy in. (Energy is not created, of course; the rest is what is absorbed at the low temperature.) Furthermore, the gain increases as the difference between the hot and cold temperatures decreases. This makes the idea of using heat pumps to heat buildings very attractive (figure 7.13). If we extract heat from a river at, say, 0 °C and deliver it to radiators at 40 °C, then, for 1 kW of electrical power consumed, an ideal heat pump would deliver nearly 8 kW of heat to the building. Heat pumps have occasionally been used for heating buildings, but, as a result of the high cost and low efficiency of practical machines, they are not generally worthwhile at present. If any reader of this book devises a really efficient and cheap heat pump, I should be delighted to share his subsequent fortune with him.

7.8 THE IDENTITY OF IDEAL GAS AND THERMODYNAMIC TEMPERATURES

We now return to the problem we had to leave in chapter 1. The formal definition of thermodynamic temperature comes out of the theory of thermal physics. In chapter 1 we stated that the temperature appearing in the equation of state of the ideal gas is thermodynamic temperature. We can now use our knowledge of ideal heat engines to prove that this is so.

We have defined thermodynamic temperature so that an ideal heat engine operating between two reservoirs exchanges heat at those reservoirs in proportion to their thermodynamic temperatures: $Q_1/Q_2 = T_1/T_2$. If we take an ideal gas around a Carnot cycle and find that it exchanges heat in proportion to temperature *as measured on its own scale*, then its own scale must be proportional to thermodynamic temperature. This is how we shall make the proof. The proof is quite long because it involves several steps, but none of them is difficult.

We start, then, by assuming that the temperature appearing in the equation of state *may* not be thermodynamic temperature, so we replace T with the usual symbol for an empirical temperature, Θ:

$$pV = R\Theta \qquad (7.11)$$

(This is equation *3.1* which defines the ideal gas scale.) What we have to show is that the heats exchanged by the gas at the reservoirs are in proportion to the Θ's of the reservoirs:

$$Q_1/Q_2 = \Theta_1/\Theta_2 \qquad (7.12)$$

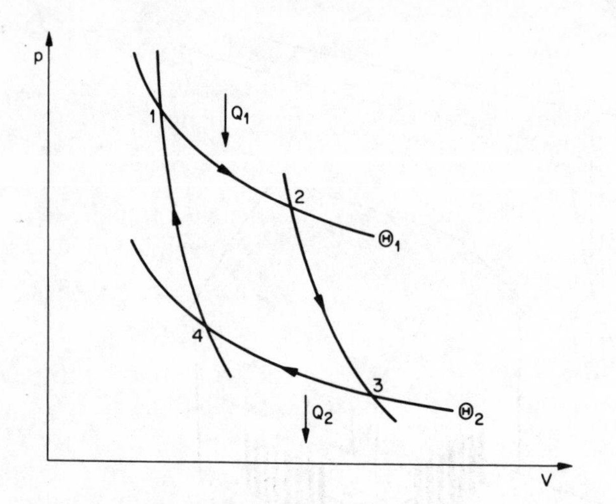

Figure 7.14 Proof that the ideal gas scale is identical with thermodynamic temperature

The Carnot cycle is illustrated in figure 7.14. $1 \rightarrow 2$ and $3 \rightarrow 4$ are the isothermal processes at Θ_1 and Θ_2. $2 \rightarrow 3$ and $4 \rightarrow 1$ are the adiabatic processes completing the cycle. We first calculate the Q's.

An ideal gas has internal energy which depends only on temperature (page 40), so the work done *on* the gas in an isothermal change must equal the heat rejected by the gas during that change (pages 104–5). Thus, the heat rejected in an isothermal change is

$$Q = -\int_{V_1}^{V_2} p\,\mathrm{d}V$$

Substituting for p with the equation of state, *7.11*,

$$Q = -R\Theta \int_{V_1}^{V_2} \frac{\mathrm{d}V}{V} = -R\Theta \ln\left(\frac{V_2}{V_1}\right)$$

Applying this to the isothermal changes in the cycle,

$$Q_1 = +R\Theta_1 \ln(V_2/V_1)$$

$$Q_2 = -R\Theta_2 \ln(V_4/V_3) = +R\Theta_2 \ln(V_3/V_4)$$

(The sign is changed in the first equation because Q_1 is heat *into* the gas. In the second case we have inverted the ratio in the ln which also changes the sign.) Hence,

$$\frac{Q_1}{Q_2} = \frac{\Theta_1 \ln(V_2/V_1)}{\Theta_2 \ln(V_3/V_4)} \qquad (7.13)$$

To arrive at *7.13* we have used the fact that $1 \rightarrow 2$ and $3 \rightarrow 4$ are isothermal changes. We have not yet used the fact that $2 \rightarrow 3$ and $4 \rightarrow 1$ are adiabatic changes.

Using pV^γ = constant for the adiabatics, we have

$$p_2V_2^\gamma = p_3V_3^\gamma$$

$$p_4V_4^\gamma = p_1V_1^\gamma$$

Using *7.11* to eliminate p from these equations,

$$\frac{R\Theta_1}{V_2}V_2^\gamma = \frac{R\Theta_2}{V_3}V_3^\gamma$$

$$\frac{R\Theta_2}{V_4}V_4^\gamma = \frac{R\Theta_1}{V_1}V_1^\gamma$$

which rearrange to give

$$\left(\frac{V_2}{V_3}\right)^{\gamma-1} = \frac{\Theta_2}{\Theta_1}$$

$$\left(\frac{V_1}{V_4}\right)^{\gamma-1} = \frac{\Theta_2}{\Theta_1}$$

Therefore

$$\frac{V_2}{V_3} = \frac{V_1}{V_4}$$

or

$$\frac{V_2}{V_1} = \frac{V_3}{V_4}$$

Hence, the two ln terms in *7.13* are equal, so that

$$\frac{Q_1}{Q_2} = \frac{\Theta_1}{\Theta_2}$$

This is equation *7.12*, the result we set out to prove. We have therefore shown that the ideal gas scale and thermodynamic temperature are identical.

7.9 ENTROPY

Our formula for the efficiency of an ideal heat engine working between two reservoirs,

$$\eta = 1 - T_2/T_1$$

shows that *heat is more useful the higher the temperature at which it is supplied.* The higher the temperature of the hot reservoir, T_1, the larger is the proportion of the heat extracted from it which is turned into useful work. The converse must also be true: the lower the temperature of a reservoir, the smaller the proportion of heat extracted that can be turned into work. We can say that *when heat flows from hotter to colder it is degraded* in the sense that less of it can be turned into work. This is not a surprising conclusion, because our analogy of water flowing downhill gives a similar result: if we let the water flow down part of the way before intercepting it, some of the potential energy is dissipated and we get less work out when the water then drives a turbine. Temperature is therefore a measure of the potential of heat for doing work.

We can illustrate this with an example. Suppose we have three reservoirs at temperatures T_1, T_2 and T_0,

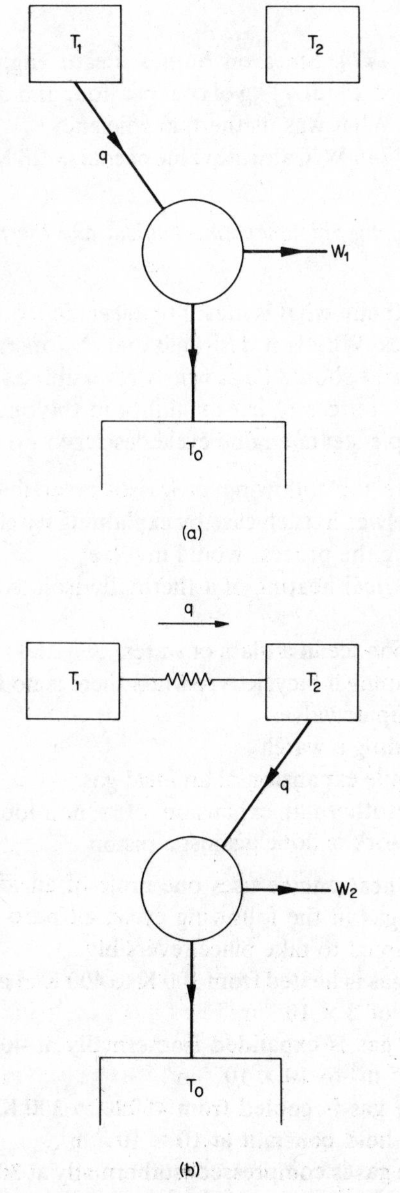

Figure 7.15 Processes illustrating the degradation of energy

T_0 being the coldest and T_1 the hottest. If we extract heat q from T_1 and use it to operate a heat engine which rejects heat to T_0 (figure 7.15a) we get work

$$W_1 = \eta q = q(1 - T_0/T_1)$$

If, instead, we first let q be conducted from T_1 to T_2 and *then* use it to operate the heat engine (figure 7.15b), the work out is

$$W_2 = \eta q = q(1 - T_0/T_2)$$

This is less than W_1 because $T_2 < T_1$. Both processes result in exactly the same changes in the energies of the hot reservoirs (the T_1 reservoir loses q and the T_2 reservoir is unchanged) but we get less work out if the heat flows to the T_2 reservoir before we intercept it to run the heat engine. The work lost is

$$W_1 - W_2 = qT_0\left(\frac{1}{T_2} - \frac{1}{T_1}\right) \qquad (7.14)$$

The loss is associated with the thermally irreversible process (page 107) of conduction from T_1 to T_2.

Now, is there a quantity which can be used to give a measure of how much irreversibility there is in a thermal process? The answer is 'yes'. The quantity used in thermodynamics is *entropy* which is defined as follows: the change in the entropy of a body at temperature T when heat Q is added reversibly to it is Q/T. The symbol used for entropy is S. Thus, in a reversible change,

$$\Delta S = Q/T \qquad (7.15)$$

In this definition, we have assumed that T does not change significantly as Q is added. In differential notation, the infinitesimal increase in entropy is given by

$$dS = dQ/T$$

If the temperature does change significantly as the heat is added, then to obtain the total entropy change, this differential expression has to be integrated taking account of the variation of T.

Entropy has the following important properties:

1) In reversible processes, total entropy is conserved. The reversible heat engine extracting Q_1 at T_1 and rejecting Q_2 at T_2 causes the entropy of the hot reservoir to change by

$$\Delta S_1 = -Q_1/T_1$$

(the entropy decreases because Q_1 is *extracted*), and

the entropy of the colder reservoir increases by

$$\Delta S_2 = Q_2/T_2$$

Therefore

$$\Delta S_{\text{total}} = \Delta S_1 + \Delta S_2 = -\frac{Q_1}{T_1} + \frac{Q_2}{T_2} = 0$$

where we have used *7.4*,

$$\frac{T_1}{T_2} = \frac{Q_1}{Q_2}$$

2) In irreversible processes, total entropy increases. When the heat q is conducted from T_1 to T_2 in figure 7.15b, the change of total entropy is

$$\Delta S_{\text{total}} = \Delta S_1 + \Delta S_2 = q\left(-\frac{1}{T_1} + \frac{1}{T_2}\right) \quad (7.16)$$

which is greater than zero because $T_2 < T_1$. Since spontaneous heat flow is always from hotter to colder, heat conduction always produces an increase in total entropy.

3) The entropy increase in irreversible processes is related to the loss of useful work which can be extracted from the system. In the illustration of figure 7.15 we have, in the irreversible process of conduction, an increase in total entropy given by *7.16*. Comparing this with the work lost, as given by *7.14*, we see

$$W_1 - W_2 = T_0 \, \Delta S,$$

so the work lost is directly related to the increase in total entropy and the lowest temperature available for working a heat engine. The generation of entropy corresponds to the degradation of the energy during the conduction process.

To discuss entropy fully is quite beyond the scope of this book; but what we have said should be enough to suggest how the properties of entropy sum up the physics of efficiency and irreversibility. In the theory of thermal physics, entropy forms the focal point of the second law. That, however, is the next part of a story which we have only begun in this book.

PROBLEMS

Things that do not happen. Kelvin and Clausius statements of the second law of thermodynamics

7.1 How does the passage of an electric current cause heating in a resistor?

Can you explain in microscopic terms (by referring to what happens to the electrons) why the reverse process does not occur: why a resistor cannot act as a heat powered battery?

Heat engines, cycles, reservoirs, efficiency

7.2 When a thermocouple with its junctions at different temperatures drives a current through a resistor, is it a heat engine?

7.3 If in one day a man eats food of calorific value 14 MJ and works for the equivalent of 8 hours at a mean rate of 0.1 hp, what is his efficiency as a heat engine?

7.4 In 1774, Smeaton built a steam engine which consumed about 42 kg of coal per hour and developed 2.65 hp. What was its thermal efficiency?
[1 hp = 746 W. Calorific value of coal = 28 MJ kg^{-1}.]

Maximizing efficiency: mechanical and thermal reversibility

7.5 Explain what is meant by *reversibility* in thermal machines. Why is it desirable that the operation of a heat engine should be as nearly reversible as possible? Identify sources of irreversibility in the operation of the simple steam engine cycle described on page 106.

7.6 Are the following processes reversible? Justify your answer in each case by explaining carefully what reversing the process would involve.
a) Electrical heating of a thermally isolated block of metal.
b) Putting ice in a glass of water.
c) Pumping a bicycle tyre when there is no friction in the pump or valve.
d) Winding a watch.
e) A Joule expansion of an ideal gas.
f) An isothermal expansion of a non-ideal gas in which work is done against a piston.

7.7 A heat engine uses one mole of an ideal monatomic gas in the following cycle, all parts of which are assumed to take place reversibly.
i) The gas is heated from 300 K to 400 K at a constant volume of 2×10^{-3} m^3.
ii) The gas is expanded isothermally at 400 K from 2×10^{-3} m^3 to 10×10^{-3} m^3.
iii) The gas is cooled from 400 K to 300 K with the volume held constant at 10×10^{-3} m^3.
iv) The gas is compressed isothermally at 300 K from 10×10^{-3} m^3 to 2×10^{-3} m^3.

a) Sketch the cycle on a p-V diagram.
b) How much heat is absorbed in (i) and (ii)?
c) How much heat is rejected in (iii) and (iv)?
d) How much work is done in the cycle?
e) What is the efficiency of the engine?
[*Hints*: remember equations *2.5*, *3.2*, *3.32* and the fact that, in an isothermal change of an ideal gas, $dQ = -dW$ (see pages 104–5).]

Carnot cycles; efficiency of Carnot engines; thermodynamic temperature

7.8 Describe with the aid of a p-V diagram a Carnot cycle of an ideal gas. How do such cycles lead to the idea of thermodynamic temperature?

7.9 The Young modulus of steel falls with increasing temperature and the metal cools if stretched adiabatically. Sketch on a graph of force against extension the form of a Carnot cycle using the stretching of a steel wire. Indicate clearly along which parts of the cycle heat is absorbed and rejected.

7.10 A reversible heat engine operating between 1500 °C and 50 °C develops 50 hp. At what rate is heat (a) absorbed at 1500 °C, (b) rejected at 50 °C?

If a real engine operating between these temperatures and delivering the same power only has 15% of the efficiency of an ideal engine, at what rate does it (c) absorb heat at 1500 °C, (d) reject heat at 50 °C?
[1 hp = 746 W.]

7.11 Coal is burnt at 1200 °C. If an ideal heat engine is powered by the burning coal, and rejects heat at 20 °C,
a) what would be its thermodynamic efficiency?
b) how much work would the engine perform per kilogram of coal burnt?
c) for how long would 1 kg of coal power a 2 kW electric fire?
d) for how long would 1 kg of coal power a 1 hp electric motor?
[1 hp = 746 W. Calorific value of coal = 28 MJ kg^{-1}.]

7.12 Two identical bodies of mass 10 kg and specific heat capacity 800 J K^{-1} kg^{-1} are initially at 20 °C. One is heated to 400 °C, and then a reversible heat engine is operated between them until they come to the same temperature.
a) How much energy is required to raise the temperature of the heated body?
b) Show that infinitesimal changes in the temperatures T_1 and T_2 of the two bodies during the operation of the heat engine are related by

$$\frac{dT_1}{T_1} + \frac{dT_2}{T_2} = 0$$

c) Hence find the final common temperature. (Remember that $d(\ln T) = dT/T$.)
d) By how much have the internal energies of the bodies changed during the operation of the heat engine?
e) How much work was obtained from the engine?
f) What proportion of the energy used in heating the body to 400 °C is recovered?
g) What happens to the rest?

Real heat engines; estimating their efficiencies. The Otto cycle.

7.13 Estimate the maximum thermodynamic efficiency of a heat engine which operates between hot cooling water from a power station (say 80 °C) and a cold river.

7.14 A solar cell converts radiation from the sun into electrical power. By treating the solar cell as a heat engine, estimate the maximum possible thermodynamic efficiency of such a device.

7.15 A 135 hp petrol engine has a compression ratio of 11.
a) What is the thermodynamic efficiency of an ideal engine using the Otto cycle and having this compression ratio?

If the efficiency of the real engine is only 40% of that of the ideal engine,
b) how long will the engine run at full power on 1 kg of petrol?
c) how long will it run at full power on 1 gallon of petrol?
d) If one gallon of petrol typically lasts half an hour, assuming the efficiency of the engine to be unchanged, what is the mean rate of working in horsepower?
[Take $\gamma = 1.4$. Calorific value of petrol = 40 MJ kg^{-1}. Density of petrol = 750 kg m^{-3}. 1 hp = 746 W. 1 gallon = 4.55×10^{-3} m^3.]

Vapour pressure; the Clausius–Clapeyron equation; approximate integrated form

7.16 Write down the Clausius–Clapeyron equation. Define the quantities appearing in the equation and give a consistent set of units for them.

An immersion heater boils water in a washing machine on a day when the atmospheric pressure is 763 mmHg. If the heater is 600 mm below the water surface, what is the temperature of the boiling water near the heater?
(Assume water vapour approximates to an ideal gas.)
[Latent heat of vaporization of water = $2.26\ \text{kJ g}^{-1}$.
Density of water = $1000\ \text{kg m}^{-3}$. Relative molecular mass of water = 18. 1 mmHg = 133 Pa.]

7.17 Use the Clausius–Clapeyron equation to find how rapidly the melting point T_m of water is lowered by application of pressure p: calculate dT_m/dp.

Use your result to estimate how much the melting point of ice under the blade of a skate is lowered by the weight of the skater. Comment on your result.
[Density of ice at 0 °C = $916.8\ \text{kg m}^{-3}$.
Density of water at 0 °C = $999.8\ \text{kg m}^{-3}$.
Latent heat of fusion = $333\ \text{J g}^{-1}$.]

7.18 Use the Clausius–Clapeyron equation to estimate the boiling point of water at the top of Everest where the atmospheric pressure is about one third of its value at sea level. Proceed as follows:
a) Use the approximate integrated form of the Clausius–Clapeyron equation (equation *7.8*) to show that the vapour pressures p_1 and p_2 at temperatures T_1 and T_2 are related by

$$\ln (p_1/p_2) = \frac{L_m}{R}\left(\frac{1}{T_2} - \frac{1}{T_1}\right)$$

b) Taking 1 to refer to boiling at normal atmospheric pressure and 2 to refer to boiling at the top of Everest, solve the equation for T_2.
[Specific latent heat of vaporization of water = 2.26 MJ kg^{-1}.
Relative molecular mass of water = 18.]

7.19 When heated, zinc sublimes (page 66). Assuming that the vapour behaves approximately like an ideal gas, show, starting from the Clausius–Clapeyron equation, that the vapour pressure p changes with temperature approximately according to the equation

$$\frac{dp}{dT} = \frac{L_m p}{RT^2}$$

where L_m is the molar latent heat of sublimation.

Integrate this equation, assuming L_m is constant, to show that the vapour pressure is given by the formula

$$\ln p = -\frac{L_m}{RT} + \text{constant}$$

Between 580 K and 630 K the vapour pressure of zinc is given by

$$\ln (p/\text{mmHg}) = -\frac{15600}{(T/\text{K})} + 20.7$$

What is the mean heat of sublimation in this range of temperature?

Refrigerators and heat pumps; efficiencies of ideal devices

7.20 If the door of a refrigerator is left open, does the room eventually cool down, warm up or stay at the same temperature? Explain your reasoning.

7.21 What is meant by the efficiency (or figure of merit) of a refrigerator? Derive an expression for the efficiency of an ideal refrigerator in terms of the temperatures between which it operates.

A domestic refrigerator operating in surroundings at 25 °C is to extract heat from an ice box at −15 °C at a rate of 10 W. What is the minimum power which would be required to drive the refrigerator?

If a real domestic refrigerator operating under these conditions has a $\frac{1}{4}$ hp motor which runs for 2 minutes each half hour, what is the ratio of its efficiency to that of an ideal refrigerator?
[1 hp = 746 W.]

7.22 An inventor claims to have developed a heat pump which will draw heat from a river at 3 °C and deliver heat to a building at 35 °C at the rate of 20 kW while consuming only 1.9 kW of electric power. Would you take his claim seriously?

7.23 An ideal domestic heating installation burns oil at temperature T_1 to power a reversible heat engine. The heat engine drives a reversible heat pump which extracts heat from the surroundings at T_3. Both heat engine and heat pump reject heat into the house at T_2. Show that the ratio of the heat delivered to the house by the complete system to that which is given directly by burning the oil is

$$\frac{T_2 (T_1 - T_3)}{T_1 (T_2 - T_3)}$$

Evaluate this ratio for T_1 = 700 °C, T_2 = 30 °C, T_3 = 0 °C.

Identity of ideal gas and thermodynamic temperatures

7.24 Show, by analyzing a Carnot cycle of an ideal gas, that ideal gas temperature (as defined by equation

3.1) and thermodynamic temperature (as defined by equation 7.4) are identical.

Entropy: conservation in reversible changes, increase in irreversible changes

7.25 Show that
a) a reversible heat engine conserves entropy,
b) entropy increases in irreversible changes,
c) an engine which violated either statement of the second law would cause entropy to decrease.

7.26 When a hot stone is dropped into a lake the change in temperature of the lake is negligible; but the stone cools down and so its entropy decreases. Is this therefore a case of an irreversible process in which total entropy decreases?

7.27 What are the changes in total entropy when
a) 20 J of heat flow from a body at 10 °C to a body at 0 °C?
b) 10 J of heat flow reversibly between two bodies at 100 °C?
c) a current of 4 A flows through a resistor of resistance 100 Ω for 3 s, the resistor being immersed in a very large volume of water at 20 °C?
d) a car weighing 1000 kg and travelling at 100 kph brakes and comes to a halt, the heat generated in the brakes being lost to the surrounding air at 20 °C?

7.28 The specific latent heats of melting and vaporization of water are 333 kJ kg^{-1} and 2.26 MJ kg^{-1} respectively. Calculate the changes of molar entropy in (a) melting of ice, and (b) boiling of water. [Relative molecular mass of water = 18.]

7.29 A reversible heat engine is operated between two bodies, one of heat capacity C_1 initially at temperature T_1 and the other of heat capacity C_2 initially at temperature T_2. As the engine operates, the warmer body gradually cools and the cooler one is warmed. By considering the changes that occur in one cycle of the engine, show that infinitesimal changes of temperature of the two bodies are related by

$$C_1 \frac{dT_1}{T_1} + C_2 \frac{dT_2}{T_2} = 0$$

Eventually, the bodies reach the same temperature T_f and the heat engine ceases to run. Show that T_f is given by

$$T_f^{(C_1+C_2)} = T_1^{C_1} T_2^{C_2}$$

What would the final temperature have been if the bodies had simply been put into thermal contact so that heat flowed (irreversibly) until they reached a common temperature?

Appendix: More Rigorous Kinetic Theory

In the account of the kinetic theory of gases in chapter 3, we did not consider in detail the random motions of the molecules. We obtained the formula for pressure by using a proof which allowed us to avoid this problem. In discussing thermal conductivity and viscosity (sections 4.2 and 4.3) we examined the physical processes taking place and deduced the *form* of the equations but could not derive the numerical values of the constants of proportionality. In this appendix, we show how the random motions are dealt with quantitatively and we give proofs of the standard results.

The problem reduces to this: we can always write down an exact result (the momentum exchanged with a surface, for example) for a molecule whose speed and direction are known. To find the total effect of all molecules we need to know how many are arriving at each speed and direction so that we may add up their individual effects to find the total.

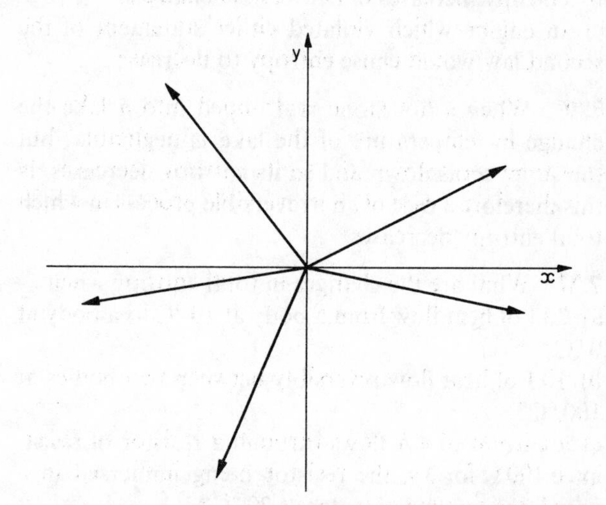

A.1 We indicate directions of motion by arrows pointing out from the origin of coordinates

A.1 DISTRIBUTION OVER DIRECTIONS

To make it easier to understand how directions are dealt with in three dimensions we will first look at the two-dimensional case: the case where all directions lie in a plane. We represent directions of motion by arrows (figure A.1). Suppose we wish to specify a range of directions. We would do this by saying that the directions lie over an angular range θ (figure A.2). But there is another way of representing this. If we draw a circle of unit radius around the origin, the range of directions corresponds to the part of the circumference through which the arrows pass. The *length* of this arc is θ, if θ is measured in radians. Thus, we measure a range of directions in a plane by *angle* defined as the length of arc on the circumference of a unit circle which is cut by the arrows (vectors) representing the directions. The unit of angular measure defined in this way is the *radian* (abbreviation,

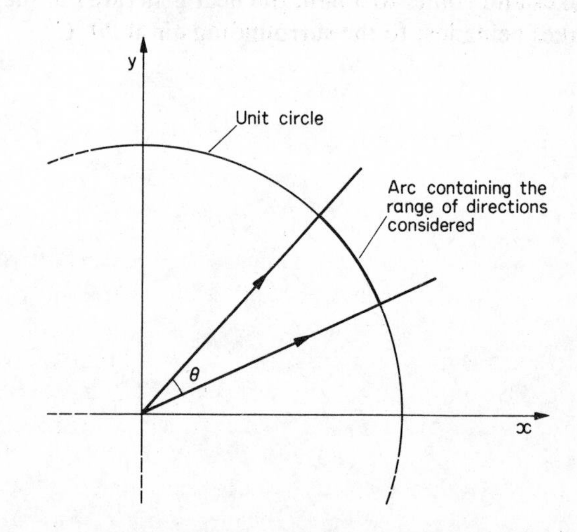

A.2 In two dimensions, a range of directions is measured by angle

rad). We note that in two dimensions, the sum of all possible directions (the whole circumference) is 2π so that the *fraction* of the whole lying in θ is $\theta/2\pi$.

We now take the three-dimensional case. The directions are again shown by arrows but this time we see how they intersect a unit sphere centred at the origin (figure A.3). A range of directions now defines an *area* on the surface of the sphere. Then, in three dimensions, a range of directions is measured by the *area* marked out on a unit *sphere*. This area is known as *solid angle* of which the unit is the *steradian* (abbreviation sr). The symbol Ω is often used for solid angle*. All possible directions correspond to the whole of the surface area of the unit sphere which is 4π. Thus a solid angle Ω is a fraction $\Omega/4\pi$ of all possible directions.

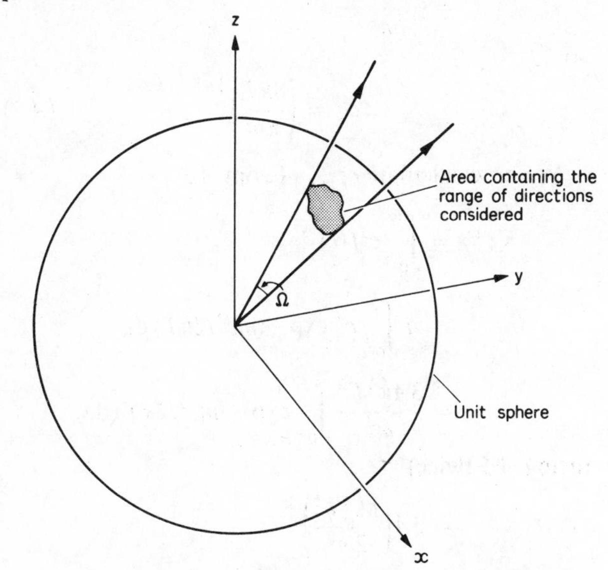

A.3 In three dimensions, a range of directions is measured by solid angle

In kinetic theory we usually only need to know the angle at which a molecule approaches a surface (i.e. the angle to the normal to the surface). A molecule arriving at 30° will exchange just as much momentum whether it approaches from the left or the right. It is therefore useful to know the solid angle which corresponds to any direction of arrival which lies between θ and $\theta + \mathrm{d}\theta$ to the normal. This is represented in figure A.4 by the shaded area. The circumference of this ring is $2\pi \sin\theta$ and the width is $\mathrm{d}\theta$ (the sphere has unit radius, remember). So the solid angle is

$$\mathrm{d}\Omega = 2\pi \sin\theta \, \mathrm{d}\theta$$

and, of the whole, it is a fraction

$$\tfrac{1}{2} \sin\theta \, \mathrm{d}\theta$$

* Ω is the capital form of the Greek letter *omega*.

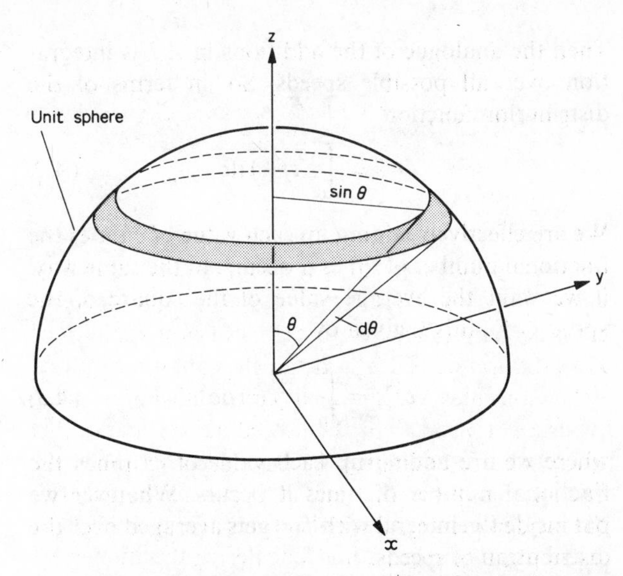

A.4 Calculation of the solid angle subtended by all directions making angles between θ and $\theta + \mathrm{d}\theta$ to a line normal to the x-y plane

A.2 DISTRIBUTION OVER SPEEDS

We have already discussed the distribution of molecular speeds (section 3.7); but we have not explained how the distribution function $f(c)$ is connected with averaging.

Suppose we have a number of boxes with various numbers of marbles in: some have one, some two, some three and so on. We calculate the average number of marbles in a box $\langle n \rangle$ with the following formula

$$\langle n \rangle = \frac{(b_1 \times 1) + (b_2 \times 2) + \ldots (b_n \times n) + \ldots}{b_1 + b_2 + \ldots + b_n + \ldots}$$

where b_n is the number of boxes with n marbles. The bottom line is just the total number of boxes. If we divide this into each of the b's on the top line this becomes

$$\langle n \rangle = (f_1 \times 1) + (f_2 \times 2) + \ldots + (f_n \times n) + \ldots \qquad (A.1)$$

where f_n is the *fraction* of the total number of boxes with n marbles.

Now in the case of molecular speeds, the possible speeds are not all whole numbers, but we do know how they are distributed because $f(c)\,\mathrm{d}c$ is defined as the fraction whose speeds are between c and $c + \mathrm{d}c$.

Then the analogue of the additions in *A.1* is integration over all possible speeds. So, in terms of the distribution function,

$$\langle c \rangle = \int_0^\infty cf(c)\,\mathrm{d}c \qquad (A.2)$$

We are effectively adding up each value of c times the fractional number of times it occurs. In the same way, if we want the average value of the square of the speed $\langle c^2 \rangle$, this is given by

$$\langle c^2 \rangle = \int_0^\infty c^2 f(c)\,\mathrm{d}c \qquad (A.3)$$

where we are adding up each value of c^2 times the fractional number of times it occurs. Whatever we put inside the integral with $f(c)$ gets averaged over the distribution of speeds.

We illustrate these procedures by calculating the relationship between $\langle c \rangle$ and $c_{\text{r.m.s.}}$.

The Maxwell distribution of molecular speeds has the form

$$f(c) = Ac^2 \exp -(mc^2/2kT)$$

where A is a constant. We first find A, using the fact that the sum of all probabilities is unity. That is

$$\int_0^\infty f(c)\,\mathrm{d}c = 1$$

Substituting for the integral,

$$\int_0^\infty f(c)\,\mathrm{d}c = A\int_0^\infty c^2 \exp -(mc^2/2kT)\,\mathrm{d}c = 1 \qquad (A.4)$$

For integrals of the form $I_n = \int_0^\infty x^n\,\mathrm{e}^{-ax^2}\,\mathrm{d}x$, there is a reduction formula:

$$I_n = \left(\frac{n-1}{2a}\right) I_{n-2} \qquad (A.5)$$

which is readily proved by integrating I_n by parts. With *A.5*, *A.4* becomes

$$\frac{AkT}{m}\int_0^\infty \exp -(mc^2/2kT)\,\mathrm{d}c = 1$$

The integral in this expression has the value $(\pi kT/2m)^{\frac{1}{2}}$, so that *A.4* becomes

$$A\sqrt{\frac{\pi}{2}}\left(\frac{kT}{m}\right)^{\frac{3}{2}} = 1$$

which gives

$$A = \sqrt{\frac{2}{\pi}}\left(\frac{m}{kT}\right)^{\frac{3}{2}} \qquad (A.6)$$

We now calculate $\langle c \rangle$. From *A.2*

$$\langle c \rangle = \int_0^\infty cf(c)\,\mathrm{d}c$$

$$= A\int_0^\infty c^3 \exp -(mc^2/2kT)\,\mathrm{d}c$$

$$= \frac{2AkT}{m}\int_0^\infty c \exp -(mc^2/2kT)\,\mathrm{d}c$$

(using *A.5*)

$$= \frac{2Ak^2T^2}{m^2}$$

so

$$\langle c \rangle = \left(\frac{8kT}{\pi m}\right)^{\frac{1}{2}} \qquad (A.7)$$

We now calculate $c_{\text{r.m.s.}}$. From *A.3*

$$\langle c^2 \rangle = \int_0^\infty c^2 f(c)\,\mathrm{d}c$$

$$= A\int_0^\infty c^4 \exp -(mc^2/2kT)\,\mathrm{d}c$$

$$= \frac{3Ak^2T^2}{m^2}\int_0^\infty \exp -(mc^2/2kT)\,\mathrm{d}c$$

(using *A.5* twice)

$$= 3A\left(\frac{\pi k^2 T^5}{2m^5}\right)^{\frac{1}{2}}$$

(substituting the value of the integral).

Substituting the value of A, we get

$$\langle c^2 \rangle = \frac{3kT}{m}$$

which is equation *3.28*. Hence,

$$c_{\text{r.m.s.}} = \sqrt{\langle c^2 \rangle} = \sqrt{\frac{3kT}{m}}$$

Comparing this with *A.7*, the expression for $\langle c \rangle$, we find

$$\langle c \rangle = \sqrt{\frac{8}{3\pi}}\,c_{\text{r.m.s.}}$$

which is the result we quoted as equation *3.30* on page 47.

A.3 FLUX

By flux of molecules we mean the number crossing unit area per second. Clearly this will also be the number per second hitting each unit area of any surface exposed to the gas. Consider molecules hitting a surface.

The total number of molecules per unit volume in the gas (the molecular density) is n. Of these,
a fraction $f(c)\,\mathrm{d}c$ is travelling between c and $c + \mathrm{d}c$
a fraction $\frac{1}{2}\sin\theta\,\mathrm{d}\theta$ is approaching the surface between θ and $\theta + \mathrm{d}\theta$.

Thus, in each unit volume there are

$$\frac{n}{2}\sin\theta\,\mathrm{d}\theta\,f(c)\,\mathrm{d}c$$

molecules which are approaching the surface between angles θ and $\theta + \mathrm{d}\theta$ and between speeds c and $c + \mathrm{d}c$. These all move towards the surface at $c\cos\theta$ (figure A.5), so the number of them which hit unit area per second is

$$\frac{n}{2}\sin\theta\,\mathrm{d}\theta\,f(c)\,\mathrm{d}c\,c\cos\theta \qquad (A.8)$$

(Number per volume × speed = number per area per second.)

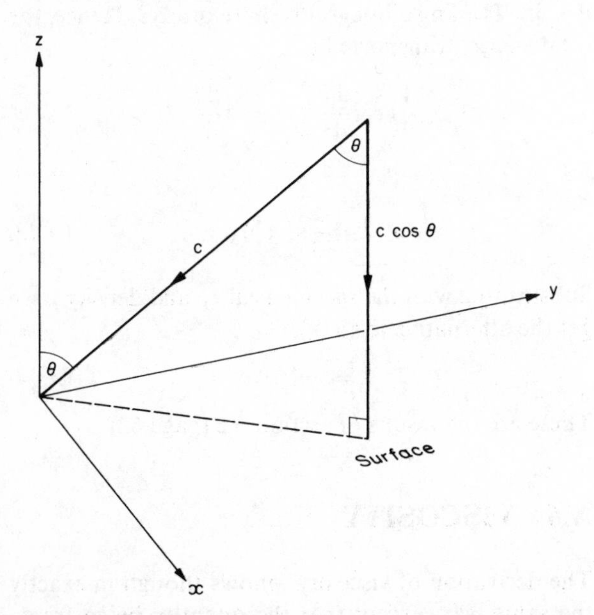

A.5 All molecules approaching a surface at speed c and angle θ to the normal approach the surface at $c\cos\theta$

The *total* number hitting unit area per second is obtained by adding up (integrating) over all possible speeds and directions:

$$\text{flux} = \frac{n}{2}\int_0^{\pi/2}\sin\theta\cos\theta\,\mathrm{d}\theta\int_0^{\infty}cf(c)\,\mathrm{d}c \qquad (A.9)$$

The angle integral is from 0 to $\pi/2$ as that includes all molecules *approaching* the surface.

$$\int_0^{\pi/2}\sin\theta\cos\theta\,\mathrm{d}\theta = \tfrac{1}{2}\int_0^{\pi/2}\sin 2\theta\,\mathrm{d}\theta$$
$$= \tfrac{1}{4}[-\cos 2\theta]_0^{\pi/2}$$
$$= \tfrac{1}{2}$$

From *A.2* the speed integral in *A.9* gives $\langle c\rangle$, so

$$\text{flux} = \tfrac{1}{4}n\langle c\rangle \qquad (A.10)$$

Using *A.7* to substitute for $\langle c\rangle$ in terms of molecular quantities, we obtain another expression for the flux:

$$\text{flux} = n(kT/2\pi m)^{\frac{1}{2}} \qquad (A.11)$$

A.4 PRESSURE

To calculate pressure one first calculates the number of molecules striking unit area per second between c and $c + \mathrm{d}c$ and between θ and $\theta + \mathrm{d}\theta$. This is *A.8* of the last section. Each of these has component of momentum normal to the surface $mc\cos\theta$ and therefore, in rebounding exchanges (normal) momentum $2mc\cos\theta$. Then the total rate of change of momentum per area per second (that is, the force per area or pressure) exerted by these molecules is

$$\frac{n}{2}\sin\theta\,\mathrm{d}\theta\,f(c)\,\mathrm{d}c\,c\cos\theta\,2mc\cos\theta$$

and the total pressure is obtained by integrating this:

$$p = nm\int_0^{\pi/2}\sin\theta\cos^2\theta\,\mathrm{d}\theta\int_0^{\infty}c^2 f(c)\,\mathrm{d}c$$

$$\int_0^{\pi/2}\sin\theta\cos^2\theta\,\mathrm{d}\theta = [-\tfrac{1}{3}\cos^3\theta]_0^{\pi/2} = \tfrac{1}{3}$$

and, from *A.3*, the speed integral is $\langle c^2\rangle$. Hence

$$p = \tfrac{1}{3}nm\langle c^2\rangle \qquad (A.12)$$

which is the result we obtained in chapter 3.

A.5 THERMAL CONDUCTIVITY

To calculate the thermal conductivity we have to find the net thermal energy carried by the molecules across a plane of constant temperature when there is a temperature gradient in the gas. Again, we first write down the energy carried by our little group of molecules which are moving between speeds c and $c + \mathrm{d}c$

and between angles θ and $\theta + d\theta$. For a molecule whose direction and speed we know, the typical energy carried can be written down exactly. We then integrate over all speeds and directions to get the total energy transported.

Suppose that temperature increases in the z direction at a rate dT/dz. We will work out the thermal energy crossing the plane $z = 0$ (figure A.6). Molecules crossing downwards will, on average, have

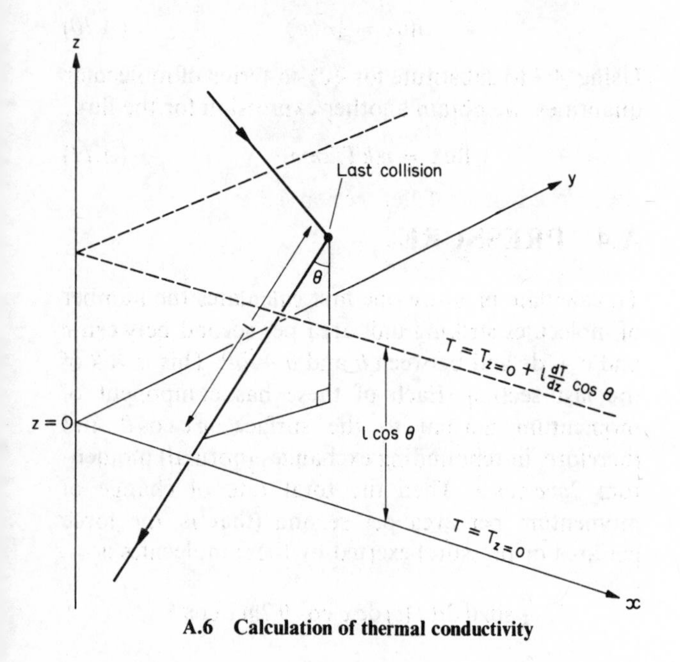

A.6 Calculation of thermal conductivity

slightly more thermal energy (they will be hotter) because their last collision took place where the temperature is higher than at $z = 0$. Similarly molecules crossing upwards will, on average, be slightly cooler. Consider one of our molecules travelling at speed c and angle θ to the normal. When it crosses $z = 0$ it will, on average, have travelled a distance equal to the mean free path l. Its last collision will therefore have been where the temperature is higher than at $z = 0$ by

$$\frac{dT}{dz} l \cos\theta$$

Its excess thermal energy will therfore be

$$c'_V \frac{dT}{dz} l \cos\theta$$

where c'_V is the heat capacity *per molecule*. The flux of such molecules crossing the plane downwards is, by *A.8*,

$$\frac{n}{2} \sin\theta \cos\theta \, d\theta \, cf(c) \, dc$$

(counting those between θ and $\theta + d\theta$ and between c and $c + dc$). So the total extra thermal energy carried by this group is

$$\frac{n}{2} c'_V l \frac{dT}{dz} \sin\theta \cos^2\theta \, d\theta \, cf(c) \, dc$$

and the net thermal energy carried downwards across $z = 0$ by all molecules is the integral of this over all directions and speeds:

$$\frac{n}{2} c'_V l \frac{dT}{dz} \int_0^{\pi} \sin\theta \cos^2\theta \, d\theta \int_0^{\infty} cf(c) \, dc$$

The speed integral is $\langle c \rangle$ and the angle integral is the same as appeared in the calculation of pressure (page 123) except that it is from 0 to π instead of 0 to $\pi/2$. This difference is because we also have to include molecules moving upwards (which will have reduced thermal energy: $(dT/dz)\cos\theta$ is negative for $\pi/2 < \theta < \pi$). The angle integral is therefore 2/3. Hence, the total energy transported is

$$\frac{1}{3} n c'_V l \frac{dT}{dz} \langle c \rangle = \lambda \frac{dT}{dz}$$

So

$$\lambda = \frac{1}{3} n l \langle c \rangle c'_V \qquad (A.13)$$

Substituting with the *specific* heat c_V and density ρ we get the alternative form

$$\lambda = \tfrac{1}{3} \rho l \langle c \rangle c_V \qquad (A.14)$$

These are the results of section 4.2 (page 62).

A.6 VISCOSITY

The derivation of viscosity follows though in exactly the same way, except that the quantity being transported is momentum.

Suppose that the speed in the x direction v_x increases with increasing z at a rate dv_x/dz. Molecules crossing $z = 0$ downwards will, on average, have slightly more horizontal momentum because their last collision took place where v_x was greater. Similarly, those crossing upwards will, on average, have slightly less horizontal momentum. Consider a molecule travelling at speed c and angle θ to the z direction. When it crosses $z = 0$ it will, on average, have travelled l since the last collision. The extra horizontal momentum it carries will therefore be

$$ml \cos \theta \frac{dv_x}{dz}$$

The flux of such molecules crossing downwards is, by *A.8*,

$$\frac{n}{2} \sin \theta \cos \theta \, d\theta \, cf(c) \, dc$$

(counting those between θ and $\theta + d\theta$ and between c and $c + dc$). So the total extra x-momentum carried by the group per second is

$$\frac{n}{2} ml \frac{dv_x}{dz} \sin \theta \cos^2 \theta \, d\theta \, cf(c) \, dc$$

and the net rate of downward transport of momentum across $z = 0$ by all molecules is

$$p_x = \frac{n}{2} ml \frac{dv_x}{dz} \int_0^\pi \sin \theta \cos^2 \theta \, d\theta \int_0^\infty cf(c) \, dc$$

Integrating, we obtain

$$p_x = \frac{1}{3} nml \frac{dv_x}{dz} \langle c \rangle$$

Hence, from the definition of viscosity, equation *4.4*,

$$\eta = \tfrac{1}{3} nm\langle c \rangle l = \tfrac{1}{3} \rho \langle c \rangle l \qquad (A.15)$$

which is equation *4.5* of section 4.3.

Useful Data

Values in these tables are given to three significant figures.

FUNDAMENTAL CONSTANTS

speed of light in a vacuum	c	$3.00 \times 10^{8}\ \mathrm{m\,s^{-1}}$
charge of a proton	e	$1.60 \times 10^{-19}\ \mathrm{C}$
Planck constant	h	$6.63 \times 10^{-34}\ \mathrm{J\,s}$
Avogadro constant	N_A	$6.02 \times 10^{23}\ \mathrm{mol^{-1}}$
unified atomic mass constant	m_u	$1.66 \times 10^{-27}\ \mathrm{kg}$
proton mass	m_p	$1.67 \times 10^{-27}\ \mathrm{kg}$
electron mass	m_e	$9.11 \times 10^{-31}\ \mathrm{kg}$
molar gas constant	R	$8.31\ \mathrm{J\,K^{-1}\,mol^{-1}}$
Boltzmann constant	k	$1.38 \times 10^{-23}\ \mathrm{J\,K^{-1}}$
Stefan–Boltzmann constant	σ	$5.67 \times 10^{-8}\ \mathrm{W\,m^{-2}\,K^{-4}}$
gravitational constant	G	$6.67 \times 10^{-11}\ \mathrm{N\,m^{2}\,kg^{-2}}$

DECIMAL MULTIPLES OF SI UNITS HAVING SPECIAL NAMES

length	ångström	$1\ \mathrm{Å} = 10^{-10}\ \mathrm{m} = 0.1\ \mathrm{nm}$
	micron (= micrometre)	$1\ \mu\mathrm{m} = 10^{-6}\ \mathrm{m}$
volume	litre	$1\ \mathrm{l} = 10^{-3}\ \mathrm{m^{3}}$
mass	tonne	$1\ \mathrm{t} = 10^{3}\ \mathrm{kg} = 1\ \mathrm{Mg}$
pressure	bar	$1\ \mathrm{bar} = 10^{5}\ \mathrm{Pa} = 0.987\ \mathrm{atm}$

c.g.s. (centimetre, gram, second) units:

force	dyne	$1\ \mathrm{dyn} = 10^{-5}\ \mathrm{N}$
energy	erg	$1\ \mathrm{erg} = 10^{-7}\ \mathrm{J}$

OTHER COMMON UNITS

mass	
pound	$1\ \mathrm{lb} = 0.454\ \mathrm{kg}$
	$1\ \mathrm{kg} = 2.20\ \mathrm{lb}$
length	
inch	$1\ \mathrm{in} = 25.4\ \mathrm{mm}$
	$1\ \mathrm{m} = 39.4\ \mathrm{in}$
volume	
pint	$1\ \mathrm{pt} = 5.68 \times 10^{-4}\ \mathrm{m^{3}} = 0.568$ litre
	1 litre = 1.76 pt
gallon	1 gall = 4.55 litres
	1 litre = 0.220 gall
pressure	
atmosphere	$1\ \mathrm{atm} = 1.01 \times 10^{5}\ \mathrm{Pa}$
millimetre of mercury	1 mmHg = 133 Pa
	1 atm = 760 × 1 mmHg
energy	
electronvolt	$1\ \mathrm{eV} = 1.60 \times 10^{-19}\ \mathrm{J}$
kilowatt-hour	1 kWh = 3.6 MJ
British thermal unit	1 BTU = 1.06 kJ
therm	1 therm = 100 000 BTU = 106 MJ
calorie	1 cal = 4.19 J
power	
horsepower	1 hp = 746 W
temperature	
Farenheit temperature	$t_F/°\mathrm{F} = 32 + \frac{9}{5}(t/°\mathrm{C})$
angle	
radian/degree	1 rad = 57.3°

OTHER DATA

s.t.p. (standard temperature and pressure) are 0 °C and 1 atm

acceleration of free fall, $g = 9.81\ \mathrm{m\,s^{-2}}$

molar volume of ideal gas at s.t.p. $= 2.24 \times 10^{-2}\ \mathrm{m^{3}}$

density of water at 4 °C $= 1000\ \mathrm{kg\,m^{-3}}$

density of air at s.t.p. $= 1.29\ \mathrm{kg\,m^{-3}}$

specific heat capacity of water at 15 °C $= 4.19\ \mathrm{kJ\,K^{-1}\,kg^{-1}}$

Answers to Problems

CHAPTER 1

1.9 (a) 69.2 °, −1.92 °, (b) 381 mm
1.10 −39.0 °
1.11 (a) 951 °, −230 °, (b) 127 Ω, 1.81 Ω
1.12 (a) 68.6 Ω, (b) 83.5 Ω, (c) 158 °
1.16 (a) 325 K, (b) 259 K, (c) 351 mm
1.17 0.24 %
1.18 (a) 257 °, (b) $a_0 = 0, a_1 = 3.92 \times 10^{-2}$ mV K^{-1}, $a_2 = 3.6 \times 10^{-5}$ mV K^{-2}, (c) 8.36 mV, (d) within about 1 K
1.19 (a) 125.8 Ω, (b) 126.0 Ω, (c) too high, (d) 0.5 K
1.20 (a) $a = 6.1, b = -0.017$ K^{-1}, (b) 0.26 K

CHAPTER 2

2.5 (a) 800 W, 1 hp, (b) 100 W, 0.1 hp
2.6 (a) 3.6 J, (b) 13.4 m s^{-1}
2.7 58.9 J, 97.3 μW
2.8 402 J
2.9 78.5 kJ, 78.5 kJ, 7.85 × 10^4 Pa, 7.85 W
2.10 10 J
2.11 $q\left(\frac{1}{2}\frac{q}{C} \pm E\right)$
2.12 60 W
2.13 (b) 10 mJ, (c) 5 mJ, (d) 5 mJ
2.19 6.28 mJ
2.23 41.7 g s^{-1} = 2.5 × 10^{-3} m^3 min^{-1}
2.24 (a) 2.89 J, (b) 4 min 49 s, (c) 2.16 K s^{-1}
2.26 (a) 400 J K^{-1}, (b) 1.5 °C min^{-1}
2.27 390 J K^{-1} kg^{-1}
2.28 $C_{ice} : L : C_{water} = 2.05:332:4.22$
2.30 55 MJ kg^{-1}
2.31 4 min 22 s, 25 min 4 s
2.32 198 J g^{-1}
2.33 0.13
2.34 3.6 × 10^{-2} m^3
2.35 0.090
2.36 389 J K^{-1} kg^{-1}
2.37 117 J K^{-1} kg^{-1}
2.38 390 J K^{-1} kg^{-1}
2.39 0.46 mm
2.40 (d) 2.43 m s^{-1}
2.41 (c) 1.01 kJ K^{-1} kg^{-1}
2.42 (a) 0.1 m^3 s^{-1}, (b) 189 kJ, (c) 23.3 kW
2.44 (a) 106 mW, (b) 46 mA
2.45 209 W m^{-1} K^{-1}
2.46 1.2 kW
2.47 24 mm
2.48 18
2.49 0.64 K
2.50 3 μm s^{-1}
2.51 (a) 13.9 °C, (b) 740 μV
2.52 (a) 219 W, (b) 97 mg s^{-1}

CHAPTER 3

3.1 2.32 × 10^{-2} m^3 mol^{-1}
3.2 763 mmHg
3.3 61.4 °C
3.4 1.21 atm
3.5 10.5 mJ
3.7 (a) 0 K, (b) 0.8 atm, (c) 2:3
3.8 (a) 0 K, (b) 3.2 atm, (c) 2:3
3.10 124 °C
3.11 9.15
3.12 (a) 94.8 K, (b) 6.1 %
3.14 $\gamma = 1.66$
3.15 (b) 4.56 × 10^{-5} m^3, (c) 3.81 J, (d) 9.22 J, (e) 13.0 J
3.16 (f) 0.0210 mol s^{-1}, (g) 579 K, (h) 262 W
3.18 0.939, 1.7
3.19 (a) 1.30 km s^{-1}, (b) 1.40
3.21 29.07 g
3.22 (a) 502 m s^{-1}, (b) 6.69 × 10^{-21} J, (c) −1.50 × 10^{-19} J
3.23 1.49 kg
3.24 2.5 × 10^{22}
3.25 (a) 4.65 × 10^{-26} kg, (b) 28 g, (c) 2.69 × 10^{19}, (d) 493 m s^{-1}
3.26 1.25 km s^{-1}

3.30 3.45 mmHg
3.31 (a) 0.21 atm, (b) 21 %
3.32 (a) 7.4×10^{-5} m^3, (b) 0.106 g
3.33 124 mmHg
3.35 5/3, 7/5, 4/3
3.36 0.9 nm
3.37 (c) 6.33 μV
3.39 (a) 2.9×10^{-31} J, (b) 2×10^{-8} K
3.40 (a) 8.57×10^{13} s^{-1}, (b) 5.65×10^{-20} J, (c) 4000 K
3.42 8.24 km, 0.30 atm
3.43 (a) 5.09×10^{14} s^{-1}, (b) 3.36×10^{-19} J = 2.10 eV, (c) 2.66×10^{-11}
3.45 (c) 490–510 m s^{-1}
3.46 (a) 2.33 km s^{-1}, (b) 7×10^{-16}

CHAPTER 4

4.2 (a) 2.68×10^{25} m^{-3}, (b) 2.04×10^{-10} m
4.3 (a) 2.11×10^{28} m^{-3}, (b) 3.62×10^{-10} m, (c) 90.4 nm
4.4 (d) 2.63×10^{21} m^{-3}, (e) 3.5×10^{-10} m
4.5 (a) 656 m s^{-1}, (b) 88 nm, (c) 2.55×10^{25} m^{-3}, (d) 0.376 nm
4.7 10.2 mm s^{-1}
4.8 1.6×10^{-5} Pa s
4.9 $A_{r,\mathrm{Ne}} = 38.6$, $S_{\mathrm{Ne}}/S_{\mathrm{He}} = 1.39$
4.12 no; yes
4.13 (a) 29.3 K, (b) 133 atm = 1.34×10^7 Pa

CHAPTER 5

5.2 (a) 1.47×10^{-28} m^3, (b) 0.26 nm (assuming cubic packing; see figure 5.9
5.3 (d) 6×10^{12} Hz, (e) 270 K
5.4 (a) 1.87×10^8 Pa, (b) 9.9×10^{-4}, (l) 1.89×10^{11} Pa
5.6 (a) 4.41 μm, (d) 2.45×10^{-6}
5.7 638 J
5.8 a few seconds per day
5.9 182 K
5.10 -162 mJ
5.11 (b) 1.9×10^{-5}, (c) ± 15%
5.12 T^{-1}
5.13 18.6 μm
5.14 (a) 0.012 %, (b) 0.09 %
5.15 190 N
5.16 (a) 300 mm^3, (b) 30 mm^3, (c) 2.7×10^{-3}, (d) 5.4×10^7 Pa = 530 atm
5.17 79 %
5.18 2.3×10^{-23} J, 0.14 meV, 1.7 K
5.19 14.3 mg s^{-1}, 20 mm^3 s^{-1}
5.20 (a) 7.8×10^{18}, (b) 1.8×10^{-23} J, 0.11 meV, 1.3 K
5.21 300 mm
5.22 2.2×10^{-2} N m^{-1}
5.23 59.5 mm
5.24 45.9 mN
5.25 5.15 mm
5.26 6.9×10^7 Pa, 680 atm
5.27 33 kJ mol^{-1}
5.28 (c) 67.4 mmHg
5.29 44 %
5.30 4.3 eV
5.31 11.7 nm
5.32 about 50 nm
5.33 380 nm
5.34 0.038

CHAPTER 6

6.2 0.46 mK
6.3 0.53 mW
6.4 7 %
6.5 33
6.8 2970 K
6.9 220 W, 185 W
6.10 (a) 13.9 kW m^{-1}, (e) 1340 K, (f) 11.5 kW
6.11 (a) 1.26 mW, (b) 5.0 nW, (c) 0.14 μW
6.12 1.5 K
6.13 (a) 1.58 kW m^{-2}, (b) 2.02×10^{17} W, (c) 289 K
6.14 92000
6.15 36 mm, 58 μm
6.17 0.43 K min^{-1}
6.18 0.027 K s^{-1}, 0.026 K s^{-1}
6.19 4.7 μW
6.20 0.5 mJ
6.21 (a) 0.725 mm, (b) 10 μm, (c) 2.9 μm, (d) 967 nm, (e) 29 pm
6.23 $\sim$3 W
6.24 (a) 6.62×10^{-34} J, 4.14×10^{-15} eV, 4.80×10^{-11} K
(b) 3.97×10^{-25} J, 2.48 μeV, 28.8 mK
(c) 5.67×10^{-20} J, 0.35 eV, 4.11 kK
(d) 1.29×10^{-15} J, 8.06 keV, 93.5 MK
6.25 (a) 3, 5 (c) 2×10^{-21} J $\approx$ 10 meV

CHAPTER 7

7.3 0.15
7.4 0.6%
7.7 (b) (i) 1.25 kJ, (ii) 5.35 kJ, (c) (iii) 1.25 kJ, (iv) 4.01 kJ, (d) 1.34 kJ, (e) 20.3%
7.10 (a) 45.6 kW, (b) 8.31 kW, (c) 304 kW, (d) 267 kW
7.11 (a) 0.801, (b) 22.4 MJ, (c) 3 h 6.9 min, (d) 8 h 20 min
7.12 (a) 3.04 MJ, (c) 171 °C, (d) −1.83 MJ, +1.21 MJ, (e) 620 kJ, (f) 20.4%
7.13 20%
7.14 95%
7.15 (a) 61.7%, (b) 98 s, (c) 5 min 35 s, (d) 25 hp
7.16 101.77 °C
7.17 -7.43×10^{-8} K Pa^{-1}
7.18 71 °C
7.19 130 kJ mol^{-1}
7.21 1.55 W, 1:8
7.22 no
7.23 7.27
7.27 (a) 5.0 mJ K^{-1}, (b) 0 J K^{-1}, (c) 16.4 J K^{-1}, (d) 1.32 kJ K^{-1}
7.28 (a) 21.9 J K^{-1} mol^{-1}, (b) 109 J K^{-1} mol^{-1}
7.29 $(C_1 T_1 + C_2 T_2)/(C_1 + C_2)$

Index